# The Journey to Quality

by

Mariwyn Tinsley and Mona G. Perdue

New View Publications
Chapel Hill

The Journey to Quality
Copyright © 1992 Mariwyn Tinsley and Mona G. Perdue

Second printing 1993

All rights reserved. Printed in the United States of America. No part of this book may be used or reproduced in any manner whatsoever without written permission except in the case of brief quotations embodied in critical articles or reviews. For information, address New View Publications, P.O. Box 3021, Chapel Hill, N.C. 27515-3021.

ISBN 0-944337-10-4

Library of Congress Catalog Card Number: 92-50328

Quantity Purchases

Companies, professional groups, clubs, and other organizations may qualify for special terms when ordering quantities of this title. For information contact the Sales Department, New View Publications, P.O. Box 3021, Chapel Hill, N.C. 27515-3021.

Manufactured in the United States of America.

## Special Thanks

♦ To **Dr. William Glasser** whose ideas will accompany us on every step of our journey which will last a lifetime.

♦ To **Perry Good** who expanded our concept of happiness, first through her book, and then by publishing ours.

♦ To **Kathy Curtiss**, our teacher in the process of learning to live the concepts of Control Theory and using the process of Reality Therapy.

♦ To **Pattye Pennachi** who never gave up on the completion of this project and provided support, encouragement, faith and ideas along the way.

♦ To **Tom** and **Ed** who have always provided support for our M&M projects through the years.

# JOURNEY TO QUALITY

## CHAPTER DISCUSSIONS

| | | |
|---|---|---|
| Introduction | | ii |
| Facilitator's Role | | iii |
| Chapter 1 | Basic Needs - Belonging | 1 |
| Chapter 2 | Basic Needs - Fun | 10 |
| Chapter 3 | Basic Needs - Freedom | 18 |
| Chapter 4 | Basic Needs - Power | 26 |
| Chapter 5 | Basic Needs - Survival | 34 |
| Chapter 6 | Agreeing How To Treat Each Other | 42 |
| Chapter 7 | Our Vision of a Quality School | 50 |
| Chapter 8 | Quality Assignments & Quality Work | 57 |
| Chapter 9 | The Expanded Role of the Teacher | 65 |
| Chapter 10 | Quality Manager: The Role of the Teacher | 73 |
| Chapter 11 | Coercion: Positive and Negative | 81 |
| Chapter 12 | Choices We Make | 90 |
| Chapter 13 | The Environment for Non-Coercion | 100 |
| Chapter 14 | Problem Solving Without Coercion | 108 |
| Chapter 15 | Using Self-Evaluation | 118 |
| Chapter 16 | Replanning the Plan | 128 |
| Chapter 17 | Problem Finding and Problem Solving: | 139 |
| Chapter 18 | Help Beyond the Classroom | 148 |
| Chapter 19 | Making a Plan for Learning | 156 |
| Chapter 20 | Continuing the Journey in a Quality School | 163 |
| References | | 171 |

# INTRODUCTION

Welcome to **The Journey To Quality**, a discussion guide that will be helpful in applying the concepts of Dr. William Glasser's Control Theory, Reality Therapy and **The Quality School** to your life personally and your life professionally.

**The Journey To Quality** is organized into staff discussion chapters which become the focus in the classroom for that week. Staff members will use the key concepts of Control Theory and Reality Therapy in their own lives before translating them from the "staff room to the classroom."

Ideally, staff members who share the same vision of what they want and a commitment to **The Quality School** will voluntarily agree to meet regularly each week to study, discuss and focus together on the concept(s). This opportunity to process Control Theory, Reality Therapy and concepts from **The Quality School** as a community of learners will be useful in developing self-understanding and preparing to teach students in the classroom. Teams of educators or study groups, as well as individuals who want quality, will find support and structure in this book as they make their own journey using the concepts of Dr. Glasser.

Recommended reading along the journey includes **The Quality School, Control Theory in the Classroom, Control Theory** and **The Quality School Training Bulletins**, all written by Dr. Glasser. Also studying and understanding Control Theory and Reality Therapy is basic for successful participation in **The Journey To Quality**.

The authors encourage you to join **The Journey to Quality**. You are invited to discover the value of thinking, learning and processing together with colleagues. Journaling and using the self-reflections and evaluation pages will be extremely helpful as you reflect on your classroom practices of the concepts. Use the book as a process guide and record any ideas or perceptions that you might want to share with others or use next year that might be helpful. Your personal observations will be helpful to others who are committed to **The Quality School** and will help all of us get more of what we want: quality in our lives and in our school.

**Mariwyn Tinsley**                                                                 **Mona G. Perdue**

# THE JOURNEY TO QUALITY
## Facilitator's Role

If you were beginning a journey to an unknown land you would want to locate a guide to help you find your way. **The Journey To Quality** is written with the idea in mind that a guide or facilitator would be helpful to each team or school taking the journey.

The facilitator's role is much like that of a travel guide. It will help you keep your travel focused on the destination. As we begin the journey we would suggest that you select a building facilitator, preferably the principal, who will serve as the guide and will help make the connections between Control Theory, Reality Therapy and **The Quality School** for the staff members as needed. Another idea would be to rotate the facilitator's role each week with a different member of the staff sharing the responsibility.

The role of the facilitator will be instrumental in establishing the learning atmosphere for the staff and seeing that a friendly, need-fulfilling environment is created at each chapter discussion.

**Facilitator's Role:**

1. Before the staff meeting, review this week's **Journey To Quality** Discussion Chapter. If the concepts are unclear, read the supplemental readings.

2. Arrange to meet weekly for one half hour with the staff members who have agreed to take the journey with you. Invite all staff members to participate. Go with volunteers. Others will come along when they are ready. Posting a sign about the meeting and the topic would be helpful.

3. All components of the discussion chapter provide a valuable opportunity for self-understanding. The self-reflection and evaluation pages are included to help each individual gain a deeper understanding of the concepts. Please do not skip components of the chapter. Each part is essential. The process of journaling can add new insight for you on your personal journey.

4. Establish staff member teams who will serve as process partners to share ideas, concerns and problems during the journey.

5. The section, **From the Staff Room to the Classroom,** is a personal process time for teachers before they work with their class. This section provides for reflection time with the major concepts taught.

6.  **The Journey To Quality** is a new beginning and may create questions, concerns or ideas. The authors invite you to share the journey with them. If you have questions or if we can be of help please call and leave a message. We'll return your call as soon as possible.

New View Publications
(800) 441-3604

## *The Journey to Quality Workshops*

Workshops and classes on implementation strategies for the Journey to Quality can be offered to staffs, districts or teams. Contact New View Publications at 1-800-441-3604 for details.

**JOURNEY TO QUALITY**

# Journey 1
# Basic Needs - Belonging

**DISCUSSION OUTCOMES AND LEARNINGS**

- to plan ways to meet our basic need for belonging responsibly
- to identify ways to create an environment where our basic need for belonging is met

Our **Journey to Quality** translates Dr. William Glasser's concepts of Control Theory, Reality Therapy and Quality School into process activities that can be used by staff members and students to gain more quality in their lives. Each discussion chapter focuses on key concepts which may be processed by staff members before classroom application. The staff discussions and feedback provide an excellent opportunity for staff and students to become learners together.

Control Theory, a theory explaining why and how we behave, is a system of how the brain operates. **Reality Therapy** is a communication process used to help individuals make more responsible choices. In Dr. Glasser's latest book, **The Quality School**, he applies three of Demings key points to education: eliminate fear and coercion, teach for quality and use self-evaluation. It is the integration of these three works by Dr. Glasser into the classroom that will lead us to quality education.

Dr. Glasser expanded William Powers' concepts of Control Theory into his book, **Control Theory**, to help explain the reasons we behave. Control Theory explains that every individual is driven by a set of five basic needs and that all behavior is internally motivated. The five basic needs which drive behavior are power, fun, freedom, love and belonging and survival. Love and belonging or connectedness is a powerful basic need that every individual has at birth. The school can help each person meet the need for belonging by creating a friendly, caring environment where every person feels involved and connected to the school. Planning how we will establish this atmosphere is an important beginning to **The Journey to Quality**. When schools create a caring climate where all people demonstrate they care about each other the school becomes a need-fulfilling place - <u>a place where everyone wants to be.</u>

**RECOMMENDED READINGS:**

Glasser, William
- **The Quality School**, Chapter 4 & 5
- **Control Theory**, Chapter 3
- **Control Theory in the Classroom**, Chapter 3

---

## ♪ Notes & "Quotes"

This space is available throughout the book for your personal notes, reflections, student observations and ideas about use of the materials. The authors are interested in any hearing about your success.

Mariwyn & Mona
New View Publications

# JOURNEY TO QUALITY

## DISCUSSION 1 WITH STAFF
## Basic Needs - Belonging

### MAKING THE CONNECTIONS TOGETHER
### ... from Concepts to Practices

1. Think about a time when you were a new person in a school, a club or in a community. Remember your feelings and what you wanted in order to feel a part of the group on your first day. **Jot down your thoughts and share them with someone sitting near you.**

2. Are there things you can <u>do</u> to feel more of a sense of belonging and connectedness when you are new? **Think first alone; then brainstorm a list together on the group chart.**

    ✎ Record on Chart: **Belonging - Things I Can Do**

3. From this list, what can we as a staff **plan** to do that will work at creating a sense of belonging in our school? **Discuss and agree on what we will do this week together that will help us move toward quality relationships.** Post the plan in the staff room in order to focus on the idea that here is a way to begin.

    ✎ Record on Chart: **Our Plan for Belonging**

4. Record the Staff Plan for Belonging here:


### WORKING IT OUT WITH STAFF

5. **Working It Out** is a hypothetical situation useful for applying the concept of the week. This week focuses on **belonging** and **responsibility**.

    > Karen comes into the staffroom on her break and finds it filled with parents who have been working on a school project. She had looked forward to a few minutes to relax and visit with her friends.

*CHAPTER 1    Page 3*

How can Karen "work it out" to meet her need for belonging **responsibly** in a way that doesn't interfere with others getting **their** needs met? **Discuss.** What are irresponsible ways Karen could meet her need for belonging? **Discuss.**

## MY PLAN FOR APPLYING THE CONCEPTS . . . Staff

6. What is one thing I plan to do this week to help me meet <u>my need</u> for belonging at school? What is something I will do to work at creating a sense of belonging with staff members? **Record on Staff Planning and Self-Evaluation page 5, now please.**

## FROM THE STAFF ROOM TO THE CLASSROOM . . .

The staff discussion parallels the classroom application this week. Using the same outcomes, you'll have the entire week to develop the concepts and integrate them into your classroom as you teach your students.

The need to develop a sense of belonging is an important beginning in any classroom. It is your first opportunity to develop involvement with the class and establish a climate of trust. This will serve as the foundation of all relationships you build in the classroom. The time it takes to do this will create connectedness for everyone.

It would be helpful for students to have guidance as they learn to plan. The guidelines for planning include it to be <u>s</u>imple, <u>a</u>ttainable, <u>m</u>easurable and <u>s</u>pecific (SAMS). We want it to be possible for students to achieve success. If the student isn't able to follow his plan, he can always make a new plan.

**Reflections, Observations and Successes** is a way of self-evaluating your plans for moving toward quality. Processing with a partner will also help you clarify your personal **Journey to Quality** each week.

♪ *Notes & "Quotes"*

## STAFF: MY PERSONAL PLAN
### For Belonging

My plan to apply the concepts personally:

What is one thing I plan to do this week to help me meet my need for belonging at school? What will I do to work at creating a sense of belonging with other staff members? **Record plan here:**

---

## STAFF SELF-EVALUATIONS AND REFLECTIONS . . .

- How well did I follow my plan for belonging at school this week?

- On a scale of 1 to 10 where am I in getting my belonging need met at school?

    1 ----------------------------- 5 ----------------------------- 10

- If I could make an improvement on what I did this week, what would I do?

---

## SUCCESSES On My Journey to Quality I Want to Share With Others:

# STAFF OBSERVATIONS 1  Basic Needs:  Belonging

1. On my journey to quality what did I do that worked well this week?

2. What would I do differently next time?

3. What is my biggest concern?  What help do I need?  What can I do?

# Belonging Ideas I Developed . . .

**Bring your journal to share at your next Journey to Quality staff discussion.**

## THIS WEEK IN THE CLASSROOM

### OUTCOMES  Basic Needs: Belonging

- to plan ways to meet our basic need for belonging responsibly
- to identify ways to create an environment where our basic need for belonging is met

### GETTING STARTED WITH STUDENTS

Encourage your students to participate in the class discussions, but be comfortable in knowing that some students may find this a new and different experience. They will join when they are ready. Let them grow in their involvement in their own way.

1. Think about a time when you were the new student at school or in a class. Remember how you felt that day and what you **wanted**. What would the teachers be doing? What would other students be doing? **Think first by yourself and then share in class together.**

2. What can you do for yourself when you are in a new school, class or situation to get what you want? **Share your ideas in the group discussion.**

3. ♪ **NOTE TO THE TEACHER:** Guide the next discussion to the basic needs everyone has. We all are born with the same basic needs. Today we are going to focus on our need for belonging. Everyone. . . your parents, teachers, fellow students, the principal, the clerk in the store . . . has the same basic need to belong, to feel a sense of connection to others. What are some things we can do in our classroom to make everyone feel a part of the class, to feel as though they belong? **Think first alone. Share ideas in class as they are recorded.**

✎ Record on Chart: **Our Need For Belonging**

4.  If this is important and we all have a need to belong, what can we **plan** to do this week in our classroom to meet our basic need for belonging? How will we treat each other? **Discuss as a class and agree on a plan for working on belonging that can be posted in the classroom.**

    ✎ Record on Chart: **Belonging: Our Plan . . .**

5.  Follow up the plan by discussing specific behaviors we **would** see and specific behaviors we **wouldn't** see if people were following the plan on how we will treat each other in order to meet our need for belonging. **Discuss specific behaviors and list on charts.**

    ✎ Record on Charts: **We Would See . . .**   **We Wouldn't See . . .**

## MY PLAN FOR APPLYING THE CONCEPTS . . . Students

6.  Explain to the students that after we've learned about our basic need for belonging, we can do something as individuals in our class to get more of what we want. Making a plan implies action and doing something. It asks for a commitment from the student, just as you've made a commitment this week.

    What is one way I can plan to get my need for belonging met in this classroom? What am I willing to do? What is my personal commitment for this week? **Share your plan with the class.**

    What can I do this week to make our classroom a place where people want to be? **Discuss.**

## APPLICATION OF LEARNING: Working It Out

7.  ♪ **NOTE TO THE TEACHER:** The problem of the week is an opportunity for students to apply what they've learned using hypothetical situations. Feel free to adapt problems to age-appropriate scenarios. You could consider using these as writing assignments, class discussions or for Cooperative Learning problem solving.

    > Jackie's best friend just moved away. What can she do to feel less lonely? How can she get a new "best" friend? What can she do to make a new friend?

    **Continue the discussion of friendship during the week.**

## STUDENT PLANNING AND SELF-EVALUATION

8. ♪ **NOTE TO THE TEACHER:** At the end of the week have students reflect in a journal, discuss with a friend or join in a class meeting/discussion as they evaluate their plans.

What did I do to meet my need for belonging today? What is one thing I did intentionally to help one other person feel like a part of our class this week?

**Additional questions for discussion during the week:**

▸ What did I do to help me meet my need for belonging today?

▸ What is one thing I did to help another person feel like a part of our class this week?

▸ How well did I do on my plan?

▸ Did I do what I said I would do?

▸ What will I do next time to get more of what I want?

♪ *Notes & "Quotes"*

# JOURNEY TO QUALITY

## Journey 2
## Basic Needs - Fun

**DISCUSSION OUTCOMES AND LEARNINGS**
- to plan ways to meet our basic need for fun responsibly
- to identify ways to create an environment where our basic need for fun is met

The need for fun is genetically programmed at birth. Every person, student and staff member has a basic need for fun everyday. In school, learning is the common ground that connects us together and provides an excellent way to meet our need for fun. Watch any young child ages 1-4 playing. The joy and excitement of learning is accomplished through play. Most of the time the teachers we remember best were teachers who created a learning environment of joy, excitement and fun.

Learning together and encouraging others to learn is fun for everyone. Part of the excellent success Cooperative Learning is having in the classroom is because it is very need-fulfilling for all. Cooperative Learning ties ways to meet the need for belonging and the need for fun together in the same activity. Without giving attention to meeting this basic need through organized learning opportunities, students may meet choose to meet their need for fun irresponsibly. Remember from your classroom days what happened when the teacher left the classroom or you had a substitute teacher. How did students meet their need for fun then?

Planning ways to add fun, laughter and play to learning will help with retaining information, keep learning exciting and meet everyone's basic need for fun. All people at every age need fun in their lives every day. When planning lessons consider all the ways you can add the element of fun to learning. It will be more need-fulfilling for both the teacher and the student.

**RECOMMENDED READINGS:**

Glasser, William
**The Quality School**, Chapter 4 & 5
**Control Theory**, Chapter 3
**Control Theory in the Classroom**, Chapter 3

# JOURNEY TO QUALITY

## DISCUSSION 2 WITH STAFF
## Basic Needs - Fun

### MAKING THE CONNECTIONS TOGETHER
### . . . from Concepts to Practices

1. Think about a time when you have a day off. What do you like to do for fun? **Contribute your ideas to a group chart.**

    ✎ Record on Chart: **Fun Away From School**

2. What do you like to do for fun in school? **Share your thoughts in group.**

    ✎ Record on Chart: **Fun At School**

3. After brainstorming, have staff members write their names by **each** of the ways they enjoy having fun. This activity will help identify interests among staff and may create the possibility for new friendships and connections.

4. If these are the things we like to do for fun, then what can we **plan** to do to insure we will have fun this week? **Discuss and work at making a staff plan for having fun that we can post in the staff room.**

    ✎ Record on Chart: **Our Plan for Fun**

5. Record the Staff Plan for Fun here:

6. Talking about the need for belonging and freedom is an important beginning in meeting basic needs in our school. How useful or helpful was the discussion last week with your students? What insights did you glean from the discussion? What did you learn? **Share with the staff.**

CHAPTER 2 Page 11

## WORKING IT OUT WITH STAFF

6. This week focuses on **fun** and **responsibility**.

    Steve and Jeff always sit at the back table during staff meetings. When they aren't talking or reading the paper they are laughing at their colleagues across the room.

    How could they meet their need for fun in a more responsible way? **Discuss**.

## MY PLAN FOR APPLYING THE CONCEPTS . . . Staff

7. What is one thing I **plan to do** to have more fun this week at school and something I **plan to do** for fun away from school? Record on Staff Planning and Self-Evaluation page, now please. See page 13.

## FROM THE STAFF ROOM TO THE CLASSROOM . . .

The staff discussion introduced the concepts you'll develop with your students this week involving the **basic need for fun and the idea of responsibility**.

The Staff Planning and Self-Evaluation and Reflection section on page 13 will provide you with the opportunity to review and record perceptions about this week and help you clarify your personal journey to quality. Please take the time to process your thoughts in this section.

♪ *Notes & "Quotes"*

## STAFF: MY PERSONAL PLAN
### For Fun

My plan to apply the concepts personally:

What is one thing **I plan to do** to have more fun this week at school and something **I plan to do** for fun away from school? **Record plan here:**

---

## STAFF SELF-EVALUATIONS AND REFLECTIONS . . .

◘ How well did I follow my plan for fun at school this week?

◘ What did I do for myself this week that was fun? How am I doing at meeting my need for fun away from school?

◘ What was the most **fun for my students to learn** this week in my classroom? What insights did I discover as I observed my students this week?

---

## SUCCESSES *On My Journey To Quality I Want to Share With Others:*

## STAFF OBSERVATIONS 2  Basic Needs:  Fun

1. On my journey to quality what did I do that worked well this week?

2. What would I do differently next time?

3. What is my biggest concern?  What help do I need?  What can I do?

## Ideas for Fun I Developed . . .

**Bring your journal to share at your next Journey To Quality staff discussion.**

## THIS WEEK IN THE CLASSROOM

### OUTCOMES Basic Needs: Fun
- to plan ways to meet our basic need for fun responsibly
- to identify ways to create an environment where our basic need for fun is met

### GETTING STARTED WITH STUDENTS

1. Think about what you like to do after school and on the weekend. What is the most fun for you? **Share with a friend and discuss with class.**

2. **Select questions from the following list to continue a discussion on the basic need for fun with your students (connect back into last week's discussion on need for belonging).**

    - What was the most fun you had at school **last year?**
    - What are ways you have fun **with a friend?**
    - What do you do to have fun when you're **alone?**
    - Is it more fun to do things with a friend or alone?
    - What do you **do** when you and your friend don't agree on what's fun?

3. Meeting our need for fun **responsibly** means that we don't interfere with others getting their need for fun met. Some people choose to meet their need for fun in a way that hurts themselves or others. Can you think of ways students meet their need for fun at the expense of others (irresponsibly)? **Discuss together in a group.** What would have been more responsible choices? **Discuss.**

4. When is **learning** fun at school? When is **learning not** fun at school? What things might get in the way of having fun at school? **Record on board or chart, if appropriate, while class discusses together.**

CHAPTER 2  Page 15

5. Does learning look and feel different when it's fun? Are there things we could do in our classroom that would make it more need-fulfilling for fun? **Discuss.**

6. Having fun is a basic human need, what can we **plan** to do every day this week in our classroom to help us meet our need for fun in a responsible way? **Discuss and make a class plan for having more fun at school that can be posted in the classroom.**

    ✎ Record on Chart: **Planning for Fun**

7. Follow up by discussing specific behaviors we **would** see and specific behaviors we **wouldn't** see if people were meeting their **need for fun responsibly. Discuss specific behaviors and list on chart(s).**

    ✎ Record on Charts: **Things We Would See**   **Things We Wouldn't See**

## MY PLAN FOR APPLYING THE CONCEPTS . . . Students

8. Explain to students that we can do some things as individuals in our class to get more of what we want. Making a plan implies action and doing something. It asks for a commitment from the student and it should be **s**imple, **a**ttainable, **m**easurable, and **s**pecific (SAMS).

    What is one thing I plan to do this week to meet my need for fun at school? What would be fun to learn this week? What are ways I can have fun in a responsible way at school (or on the playground) with my friends? **Discuss.**

## APPLICATION OF LEARNING: Working It Out

9. Jerry doesn't want to go out to recess (or be in the hallway during class breaks). What could be Jerry's problem? Have you ever felt this way? What could Jerry do? **Discuss.**

♪ *Notes & "Quotes"*

## STUDENT PLANNING AND SELF-EVALUATION

10. Continue to have students work on making plans and self-evaluating. You may want to tell them one thing **you** did in your plan as a way of modeling the self-evaluation process.

    ▸ What did I do to meet my need for fun this week?
    ▸ Was my plan successful for me?
    ▸ What is something I did alone for fun this week?
    ▸ What did I do with another person to meet my need for fun responsibly?
    ▸ What plan could I make for getting even more fun at school every day?

♪ *Notes & "Quotes"*

# JOURNEY TO QUALITY

## Journey 3
## Basic Needs - Freedom

### DISCUSSION OUTCOMES AND LEARNINGS

- to describe ways to meet our basic need for freedom
- to identify ways to create an environment where our basic need for freedom is met
- to work at planning for meeting our basic need for freedom

Freedom is a need so strong that people have fought and died for this right. Freedom is one of our basic needs that may sometime be in conflict with other basic needs. We want to be free to do what we want, but we also want to belong. Learning to give and take in relationships is learning to meet our need for freedom responsibly.

Classrooms and schools can be organized to provide ways students can meet their need for freedom responsibly. In the classroom, students can move around without permission, select specific activities of interest to do, choose reading materials, research and writing topics, and express themselves through art, drama, music or poetry. In fact, being able to select our own friends can meet a need for freedom. Freedom needs are basic for both teachers and students. In the classroom finding the balance between acceptable and unacceptable freedom or choice is something students and teachers will need to work out together.

**RECOMMENDED READINGS:**

Glasser, William
**The Quality School**, Chapter 4
**Control Theory**, Chapter 2
**Control Theory in the Classroom**, Chapter 3

# JOURNEY TO QUALITY

## DISCUSSION 3 WITH STAFF
## Basic Needs - Freedom

### MAKING THE CONNECTIONS TOGETHER
#### . . . from Concepts to Practices

1. Think about a really busy day at school and how you feel when you don't have a moment to yourself. What do you do when it seems as if you have little freedom or choice in your day? **Jot down your actions.**

2. What things do students often do when they feel as if they have few opportunities for freedom or making choices? **Share your thoughts in small groups.**

3. We all have a need for freedom and our actions are always our best attempt at the time to meet our needs. What are ways our staff could experience more freedom and choices at school? What could we do to **create an environment** where our need for more personal freedom could be met in a responsible way? **Brainstorm in the large group and record on a chart.**

    ✎ Record on Chart: **Ways to Get More Freedom**

4. From the list, what would we be willing to **do** this week that will help us get more personal freedom at school? **Discuss options and agree on a plan that will increase our freedom needs this week at school. Chart and post the plan in the staff room.**

    ✎ Record on Chart: **Our Plan to Increase Freedom**

5. Record the Staff Plan for Freedom here:


6. Now think about the opportunities for freedom and choice students have in your class. What can we do to meet **our freedom need as a teacher** and **help students** meet their need for freedom at the **same** time? What is your greatest concern? **Share ideas in large group.**

---

CHAPTER 3    Page 19

## WORKING IT OUT WITH STAFF

7. **Working It Out focuses on freedom and responsibility this week.**

    Once a month the staff enjoys going out to dinner together on payday. Several staff members have chosen **never** to attend. Peggy, a teacher, decides she wants **everyone** there this month. She makes the point at the staff meeting, singling those out who don't usually come by saying that **everyone** would have more fun if **everyone** attended.

How can all staff members meet their own need for freedom responsibly? **Discuss.**

## MY PLAN FOR APPLYING THE CONCEPTS . . . Staff

8. As you work with your students through the week, think of ways you could create more opportunities for students to experience freedom and make responsible choices. What is possible this week?

    What will I do this week to increase opportunities for my students to meet their need for freedom responsibly?

**Record on Staff Planning and Self-Evaluation page 21, now please.**

## FROM THE STAFFROOM TO THE CLASSROOM . . .

This week as you focus on freedom and choice with your students, continue guiding the evaluation of the weekly plan. Link discussions back to the basic needs of belonging and fun when appropriate.

Discussion questions can be managed and adapted in creative ways to suit your style and classroom organization. You might use Cooperative Learning groups, pairs of students working together, creative writing assignments, and problem-solving applications for processing the concept of freedom throughout the week.

*♪ Notes & "Quotes"*

## STAFF: MY PERSONAL PLAN For Freedom

My plan to apply the concepts personally:

What will I do this week to increase opportunities for my students to meet their need for freedom responsibly? **Record plan here:**


## STAFF SELF-EVALUATIONS AND REFLECTIONS . . .

◘ Was I able to get more of my personal need for freedom met this week at school? What worked? What didn't? What will I do next?


◘ Did my plan work for bringing more choice and freedom into my classroom this week? What do I plan to do next?


◘ On a scale of 1 to 10 where am I in getting **my** need for freedom met at school?

1 -------------------------- 5 -------------------------- 10

## SUCCESSES On My Journey to Quality I Want to Share With Others:

## STAFF OBSERVATIONS 3  Basic Needs:  Freedom

1. On my journey to quality what did I do that worked well this week?

2. What would I do differently next time?

3. What is my biggest concern?  What help do I need?  What can I do?

## Freedom Ideas I Developed . . .

**Bring your journal to share at your next Journey to Quality staff discussion.**

# THIS WEEK IN THE CLASSROOM

## OUTCOMES  Basic Needs: Freedom

- ◼ to describe ways to meet our basic need for freedom
- ◼ to identify ways to create an environment where our basic need for freedom is met
- ◼ to work at planning for meeting our basic need for freedom

## GETTING STARTED WITH STUDENTS

1. Think about the word freedom. It has many different meanings. What does it mean to you? **Discuss in large group.**

2. All people have a basic need for freedom. How they get their freedom need met is unique to them. What are we doing in our class when we experience a feeling of freedom? **Discuss and chart.**

   ✎ Record on Chart: **Ways We Meet Freedom Needs**

3. Why is it, in our classroom that sometimes it is okay to move around the room freely and at other times it is **not** okay? **Discuss in large group.**

4. Another word for freedom is choice. **Discuss together as a group:**

   ▸ What are some choices you made today before you left for school?
   ▸ Who is responsible for the choices you made?
   ▸ When you arrived at school, what choices did you make?
   ▸ Who is responsible for them?

5. We are responsible for our own choices and behaviors. Sometimes people make choices that could be harmful to themselves and/or others. **Briefly discuss choices we make in the classroom, and on the playground.**

CHAPTER 3   Page 23

6. Let's brainstorm ways we could get more freedom and have more choices at school that wouldn't be harmful to ourselves or others. **Discuss and decide together if the suggestions are responsible or irresponsible.** Ask: Does this get in the way of anyone else's need for freedom? **Chart.**

    ✎ Record on Chart: **Ways We Could Get More Freedom**

7. What one thing from our list can we plan to do as a group this week that will help us meet our need for freedom? **Discuss responsible choices and make a plan for meeting our freedom need that can be posted in the classroom. Chart.**

    ✎ Record on Chart: **Freedom: Our Plan . . .**

8. Follow up the plan by discussing specific behaviors we **would** see and **wouldn't** see when people are following the plan. Using the terms "helpful" and "hurtful" behavior may add clarity for younger students. **Discuss and chart.**

    ✎ Record on Chart: **We Would See. . .    We Wouldn't See . . .**

## MY PLAN FOR APPLYING THE CONCEPTS . . . Students

9. Remind students about making plans that are **s**imple, **a**chievable, **m**easurable and **s**pecific (SAMS). They've made a plan, as a group, for freedom. Now students have an opportunity to think and plan **individually** to do one thing to help meet their need for freedom.

    What is one thing I plan to do this week to meet my need for freedom in a responsible way? How can I meet my need for freedom responsibly? What am I choosing to do this week? **Share plans.**

## APPLICATION OF LEARNING: Working It Out

10. Dick is always late coming in to class. What need could Dick be meeting by choosing to come late to class? **(Please adapt the problem to your classroom, staying with the need for freedom.) Discuss together.**

# STUDENT PLANNING AND SELF-EVALUATION

11. **Invite students to reflect in a journal, discuss with a friend or join in a class meeting/discussion as they evaluate their plans.**

What did I choose to do to meet my need for freedom this week in our class?

- How well did the plan work for me?

- Did I do what I **said** I would do?

- What do I want to plan to work on now to get more freedom?

**Continue developing the idea of self-evaluation. This will lead students to make more responsible choices for themselves.**

♫ *Notes & "Quotes"*

# JOURNEY TO QUALITY

## Journey 4
## Basic Needs - Power

**DISCUSSION OUTCOMES AND LEARNINGS**
- to describe ways to meet our basic need for power
- to identify ways to create an environment where our need for power is met
- to work at planning for meeting our need for power responsibly

Power can be a difficult concept to understand because of the preconceived meaning of the word. Control Theory requires us to rethink the concept of power from "power over" people to developing a "power within." Control Theory teaches us that all our behavior is internally driven by our basic needs. The need for power is common for all human beings. Everyone has a need to be competent, to have people respect them and feel a sense of importance in what they do. In fact, we often compare ourselves to others just to see how our power need measures up. By itself power is neither good nor bad. It's how we meet our need for power that is the critical point.

When we accomplish something difficult, we experience a feeling of power, competence and worth within ourselves. For success in school students must experience power in their academic classes. Without this, students may choose to meet their need for power on the playground, in the halls, or after school by using "power over" another students.

It is important that schools be need-fulfilling by creating conditions where **everyone experiences success everyday**. Students and teachers need successful learning experiences in a supportive environment where they are free to risk, to try and to achieve something everyday and experience their power needs being met responsibly.

RECOMMENDED READINGS:

Glasser, William
**The Quality School**, Chapter 4 & 5
**Control Theory**, Chapter 3
**Control Theory in the Classroom**, Chapter 3

# JOURNEY TO QUALITY

## DISCUSSION 4 WITH STAFF
## Basic Needs - Power

### MAKING THE CONNECTIONS TOGETHER
### ... from Concepts to Practices

1. Think about something that was very difficult for you to learn but you persevered until you finally accomplished it. What words would you use to describe your behaviors when you mastered the skill? What were you feeling? What were you thinking? What were you doing and how did your body respond? (Control Theory suggests we always behave as a Total Behavior System.) **Discuss.**

2. When we accomplish something that is difficult we gain a sense of power or competence and our need is met **responsibly.** We gain an internal power of ourselves, not external power over another person. What are some conditions we can create in our classrooms to help students gain a sense of power responsibly. **Discuss and record on chart.**

    ✎ Record on Chart: **Success Conditions for Power**

3. Think of an example from your classroom of a student who is meeting the need for power in an irresponsible way. Control Theory says that the student is using the best behavior he/she knows at the time. Considering what we know about basic needs, how can we influence the student to make more responsible choices. **Discuss and brainstorm ways to help students become more responsible.**

4. One way of meeting the need for power is to **empower others.** What are ways we can empower others? **Share your examples with another staff member.**

5. Knowing what we know about power, what are ways we as staff members can get our need for power met in our school. **Think alone, then brainstorm a list in the large group. Chart ideas.**

    ✎ Record on Chart: **Power: Things We Can Do**

CHAPTER 4    Page 27

6. Discuss ideas from the list that could serve as a **s**imple, **a**ttainable, **m**easurable and **s**pecific (SAMS) plan for getting **more** of our need for power met responsibly at school this week. **Agree on a plan. Chart.** Post the plan in the staff room.

    ✎ Record on Chart: **Our Plan for Power**

7. Write the Staff Plan for Power here:

## WORKING IT OUT WITH STAFF

8. Sarah, a teacher nearing retirement, has been in the same building for 20 years. She likes to take all new teachers "under her wings" and tell them how the school operates and exactly what they are expected to do. She rules the school from her 3rd grade classroom.

What can Sarah do to meet her need for power more responsibly. **Discuss and share ideas.**

## MY PLAN FOR APPLYING THE CONCEPTS ...Staff

9. Think of ways you get your need for power met at school. Some ways our power need can be met are through the feeling of competence we experience when we have done a difficult task, experiencing a sense of importance when we accomplish a goal, or the feeling we experience when we are able to help others. Who determines if our need for power is met? What is one thing I plan to do this week that will give me a feeling of power? **Record on Staff Planning and Self-Evaluation page 29, now please.**

## FROM THE STAFFROOM TO THE CLASSROOM . . .

Dr. William Glasser says that meeting our need for power can be difficult. We tend to think of our power need at school as being met through academic success. What about a student who experiences little academic success at school? Since power is a need we all have, how will the academically deficient students get their needs met at school? What are ways all students can meet their power needs in a responsible way? Everyone needs to be good at something, not necessarily something academic.

Remember to include your successes on the journal self-evaluation page. We can all benefit from each other's ideas with this challenging concept.

CHAPTER 4    Page 28

## STAFF: MY PERSONAL PLAN for Power

My plan to apply the concepts personally:

What is the one thing I plan to do this week to meet my need for power at school?
**Record plan here:**

## REFLECTIONS . . .

◘ Was I able to get more of my personal need for power met this week at school? How well did I follow my plan? What will I do next week?

◘ How do I know if my students understand the Control Theory concept of responsible power?

◘ On a scale of 1 to 10 where am I in getting my need for power met at school?

1 ------------------------ 5 ------------------------ 10

## *SUCCESSES On My Journey to Quality I Want to Share With Others:*

# STAFF OBSERVATIONS 4   Basic Needs:  Power

1. On my journey to quality what did I do that worked well this week?

2. What would I do differently next time?

3. What is my biggest concern?  What help do I need?

## Ideas for Power I Developed . . .

**Bring your journal to share at your next Journey to Quality staff discussion.**

# THIS WEEK IN THE CLASSROOM

## OUTCOMES Basic Needs: Power

- to describe ways to meet our basic need for power
- to identify ways to create an environment where our need for power is met
- to work at planning for meeting our need for power responsibly

## GETTING STARTED WITH STUDENTS

1. Think about something you know how to do this year that you didn't know how to do last year. What is it? Describe how you felt when you learned how to do it. What you were thinking and doing. Describe your feelings when you were learning? **Share in class discussion.**

2. When we learn something that is hard for us we feel a sense of accomplishment or power. This is meeting our power need responsibly. Power is competence and knowledge. It is a feeling **inside** of us of internal power and strength. When we try to have **power over** people we are using force (external power). This is an irresponsible use of power. In our school and classroom we want to meet our need for power responsibly. What are some words we can use to describe what we mean by **responsible** power? **Ask students.** (Other words describing power are: accomplishment, competence, recognition, and gaining importance.)

3. What are things you do at home or school that help you meet your need for power? **Discuss with students and generate two lists on chart paper or the chalkboard.**

    Record on Charts: **Power at Home**   **Power at School**

4. If we had our ideal classroom environment where everyone's need for power was met, what would it look like?

CHAPTER 4   Page 31

- What would people be doing?
- What would people be thinking?
- What would people be feeling?

**Discussion of these questions may help the students envision a classroom where everyone feels important, validated and competent.**

5. Now that we have a picture in our minds of what we want, what can we plan to do **every day** in our class to help us meet our need for power in a way that is responsible? **Discuss and agree on a plan of the week that can be posted in the classroom.**

    ✎ Record on Chart: Our Plan This Week . . .

6. Follow up the plan by discussing specific behaviors we **would** see and **wouldn't** see in our school if people were following the plan on how we will **meet our need for power responsibly. Discuss specific behaviors and list on charts.**

✎ Record on Chart: What We Would See . . .    What We Wouldn't See . . .

## MY PLAN FOR APPLYING THE CONCEPTS
### . . . Students

7. What is **one** way I can plan to get my own need for power met every day this week? **Share your plan with the class.**

8. What can I do this week so that I won't interfere with others getting their need for power met?

## APPLICATION OF LEARNING: Working It Out

9. Chris is the class clown. Every time the class is enjoying a film, discussion or independent work time he gets silly and acts out. How can Chris get his need for power met in a more responsible way? **Discuss together. Create additional problems for your class to solve.**

# STUDENT PLANNING AND SELF-EVALUATION

10. Self-evaluation is an on-going process to use with students.

    **Younger Students**
    Draw a picture and/or write about something you accomplished this day/week that met your need for power.

    **Grades 3-12**
    Describe what you did well this week. Write a journal entry on something you did well this week at school. Sometimes people put a letter "Q" on their very best work. "Q" stands for quality. What school work did you do this week that you could put a "Q" on?

♫ *Notes & "Quotes"*

# JOURNEY TO QUALITY

## Journey 5
## Basic Needs - Survival

**DISCUSSION OUTCOMES AND LEARNINGS**

- to identify ways to create an environment where our basic need for survival is met responsibly
- to learn that signals are felt as an urge to behave
- to understand that we make our own choices for how we behave

Survival is a physiological need which is part of the reptilian (old) brain system and its function is reproduction and surviving. This basic need can best be described when we think of our need for food, clothing, air, water, shelter, and safety.

If our basic need for survival is not met, our body sends us a frustration signal, or urge to behave. When the signal is strong enough, we act to reduce the signal and get more of what we want. If we are hungry we will experience a signal from our body that says, "I'm hungry." We will do something (eat) to reduce the signal. Other body signals we may receive are: I'm cold, I'm hot, I'm uncomfortable. The physiological need for survival must be met before the four psychological needs (fun, freedom, power and belonging) can be met.

To reduce the frustration signal we must act or do something differently. We always behave as a Total Behavior System. Our behavior system includes four behaviors: physiology (body), feeling, thinking and doing. All of our behaviors always operate together creating our Total Behavior System and we behave to meet our basic needs.

**RECOMMENDED READINGS:**

Glasser, William
The Quality School, Chapter 4
Control Theory, Chapter 2 & 4
Control Theory in the Classroom, Chapter 2 & 3

# JOURNEY TO QUALITY

## DISCUSSION 5 WITH STAFF
## Basic Needs - Survival

### MAKING THE CONNECTIONS TOGETHER
### . . . from Concepts to Practices

1. Think about some of the workshops you have attended or about one you have been anticipating because of your interest in the content. You arrive at the meeting and there is no coffee or snacks. The room is small and crowded and you can't see the instructor or read the overhead projector. Even with the desire for knowledge you are unable to focus on the content because of your physical needs. **What could you do to solve the feelings you are experiencing? What basic need is not being met? Have several staff members share their experiences.**

2. Our survival need must be satisfied before our four psychological needs can be met. In our building are there survival needs that people are experiencing that are getting in the way? Examples could include lack of staff room seating for lunch, not enough restrooms, limited number of telephones, someone's lunch being taken, etc. What could staff members do to reduce their <u>frustration signals</u> and get their survival need met? **Discuss and share ideas with others.**

3. Considering all the physical/survival needs that staff members have identified, are there things we as a staff, could agree to **do** that would help each of us get our survival need met more responsibly? **Discuss and come to an agreement.**

   ✎ Record on Chart: **Survival: We Agree To . . .**

### WORKING IT OUT WITH STAFF

4. It's a warm spring afternoon in your classroom. The students are hot and you have no air conditioning or way to cool the room. They start to complain about the heat. You sense they are frustrated.

CHAPTER 5   Page 35

What are some ways students could reduce the signals they are experiencing and change their **Total Behavior** (physiology, feelings, thinking and doing)? **Discuss, share ways students could reduce the frustration signal they are experiencing.**

## MY PLAN FOR APPLYING THE CONCEPTS
### . . . Staff

5. What is an unsatisfied survival need I am frustrating over at school? **What will I do to reduce the frustration?** What is my plan? Record on Staff Planning and Self-Evaluation page 37, now please.

## FROM THE STAFFROOM TO THE CLASSROOM . . .

During the class discussions this week it is a good time to make the connection to all five basic needs. If the **survival need** is met, then the **four psychological needs**, (belonging, fun, freedom, power) can be considered. Another way of explaining survival needs to students is to refer to the need as our **body wheel**.

♫ *Notes & "Quotes"*

# STAFF: MY PERSONAL PLAN For Survival

My plan to apply the concepts personally:

What is one thing I plan to do to meet my survival need this week at school? **Record plan here:**

## STAFF SELF-EVALUATIONS AND REFLECTIONS . . .

◼ How well did I follow my plan to meet my survival need at school?

◼ Am I recognizing signals my body is sending me? Give an example.

◼ What else can I do to meet all my basic needs at school?

## SUCCESSES On My Journey to Quality I Want to Share With Others:

## STAFF OBSERVATIONS 5  Basic Needs:  Survival

1. On my journey to quality what did I do that worked well this week?

2. What would I do differently next time?

3. What is my biggest concern?  What help do I need?

## Ideas for Meeting Survival Needs I Developed . . .

Bring your journal to share at our next Journey To Quality staff discussion.

# THIS WEEK IN THE CLASSROOM

## OUTCOMES  Basic Needs: Survival

- to identify ways to create an environment where our basic need for survival is met responsibly
- to learn that signals are felt as an urge to behave
- to understand that we make our own choices for how we behave

♪ **NOTE TO TEACHER** - Here is additional information on signals which may help you in your discussion with students.

Everyone has physical needs. Your body signals you when it needs something. We've been talking about basic needs everyone has for **fun, freedom, belonging, and power**. The most basic need your body has is to stay alive or survive. When your survival need is unmet you get signals which send a message to your brain felt as an urge to behave. When the signal is strong enough we act. We **choose** our own behavior to satisfy our need. This reduces the signal just as eating reduces the hunger signal, or putting on a sweater reduces the cold signal.

## GETTING STARTED WITH STUDENTS

1. Think about being really hungry. How do you know when you are hungry? What signal does your body send to you? Think of a time when you were really scared. Describe it. **Share in large group.**

2. **Prepare 3x5 index cards** with survival situations. Topics may include:

   ▸ classroom is too hot
   ▸ a fierce dog approaches you
   ▸ standing and giving a talk before the class
   ▸ slamming your finger in the car door
   ▸ your cat can't get down from the tree and you're afraid of heights
   ▸ it's freezing cold outside

CHAPTER 5    Page 39

Divide students into Cooperative Learning groups. Each group draws a situation card and identifies the signal the body would feel in this situation. What could the person do to reduce the signal and get their survival need met? **Each group presents to the large group.**

3. Being **safe** is also a survival need we all have. What can I do to be safe at school both in and out of the classroom? What are some responsible safety choices I will make daily? (Examples may include walking vs running, leaning back in chairs, shoving, pushing, tripping, etc.). **Discuss with the class and identify responsible safety choices we make in our classroom and in our building every day.**

✎ Record on Chart: **Safety Choices**

4. What are **survival needs** we have every day in our classroom that need to be met in a responsible way? List on a chart or chalkboard. (Items may include using the restroom, getting a drink, temperature of the room, being hungry, classroom is crowded and not enough personal space.) What are survival needs we need to consider in our classroom? **Discuss with the class and identify responsible survival choices we make daily.**

✎ Record on Chart: **Survival Choices**

5. What can we **plan** to do to meet our safety and survival needs responsibly every day in our classroom and in our school? **Discuss and make a class plan** for meeting our safety and survival needs that can be posted in our classroom.

✎ Record on Chart: **Planning for Safety and Survival**

♪ *Notes & "Quotes"*

## MY PLAN FOR APPLYING THE CONCEPTS . . . Students

6. Interview someone in your home and ask them to talk to you about what it was like when they went to school. What were ways their safety and survival needs were met when they were in school? What were some class "rules" or plans that were different from ours today? What plans were similar to ours? **Share in class discussions.**

7. What is one thing I **plan** to do this week to meet my survival and safety needs at school? What will I have to do differently? Are my choices **responsible? Discuss with classmates.**

## APPLICATION OF LEARNING: Working It Out

8. ♪ **NOTE TO THE TEACHER:** Please feel free to adapt the problem to fit your students: staying with the concept of survival.

   Sue often oversleeps and skips breakfast in order to catch the bus to school. What survival signal might she feel during the morning? What choices could she make to reduce the signal? What is a more responsible choice she could make? **Discuss in a small group. Share with the large group.**

## STUDENT PLANNING AND SELF-EVALUATION

9. This opportunity for self-evaluation will help students see how responsibly they are making choices.

   ▸ What did I **do** to meet my survival and safety needs this week?

   ▸ Was my **plan** successful?

   ▸ Is there anything I will **do differently** next time?

---

*♪ Notes & "Quotes"*

---

*CHAPTER 5   Page 41*

# JOURNEY TO QUALITY

## Journey 6
## Agreeing How to Treat Each Other

### DISCUSSION OUTCOMES AND LEARNINGS

- to review the five basic needs
- to combine our class plans for meeting basic needs into **one agreement** for responsible behavior
- to introduce Dr. William Glasser, author of <u>The Quality School</u>, <u>Control Theory</u>, and <u>Control Theory in the Classroom</u>

As we develop our quality classroom we will want to create a need-satisfying environment where everyone can get his or her needs met in a responsible manner. All of us are driven by the same five basic needs, but the way in which we meet them is unique to each of us. Agreeing on a plan for how we will do this in our school and in our classroom is an important step in establishing quality. Dr. Glasser believes that schools can operate with one agreement, **"Be courteous."** He believes other rules may not be necessary. Once students understand how their basic needs drive their behaviors this one rule can help us create an environment where courtesy prevails.

When a student misbehaves we have often believed it was something personal and directed toward or against the teacher. As we understand Control Theory we've learned that our behavior is always our best attempt at that moment to meet our basic needs. So a misbehaving student is using the best actions he can right then to get what he wants. As teachers, we want to learn how to help students find more appropriate, courteous, responsible behaviors that are more fulfilling to them.

**RECOMMENDED READINGS:**

**Glasser, William**
**The Quality School**, Chapter 9
**Control Theory**, Chapter 1, 2 & 3
**Control Theory in the Classroom**, Chapter 3 & 4

# JOURNEY TO QUALITY

## DISCUSSION 6 WITH STAFF
## Agreeing How to Treat Each Other

### MAKING THE CONNECTIONS TOGETHER
### . . . from Concepts to Practices

1. Think about your favorite thing to do when you have all the time that you need. What do you like to do best? Does it meet more than one basic need?

2. When do you feel the most happiness at school? Assuming that your survival need has been met, what psychological needs are being met? **Briefly share with the group what you are doing when you are the happiest at school.**

3. Dr. William Glasser writes in Chapter 9, <u>The Quality School</u>, that if students are courteous in the way they treat each other, other rules will not be needed. What does being courteous look like among staff members? **Discuss specific behaviors and list on chart.**

    ✎ Record on Chart: **Being Courteous**
    **What We Would See... What We Wouldn't See...**

4. How can we apply the idea of being courteous to all school employees? **Discuss.**

5. Let's take another look at the plans for meeting our basic needs that we have developed as a staff over the last few weeks. Could **"Be courteous"** replace our five basic need plans? **Discuss and come to an agreement.**

    ✎ Record on Chart: **How We Will Treat Others . . .**

6. Write our new staff plan here:

7. How will having a staff plan move us along on our journey to quality? What will we be doing differently? What will be the same? **Discuss together.**

## WORKING IT OUT WITH STAFF

8. Mary Ellen is a "veteran teacher" and has taught in her building for over ten years. She has staked out ownership of the AV equipment, her seat in the staff room, and her chair at the faculty meetings. Marilyn, a new teacher to the building, does not know about Mary Ellen's "territory." Marilyn comes into the staff room and sits down in "Mary Ellen's chair." Mary Ellen enters the staff room with her lunch in hand and sees Marilyn in "her" seat. She glares at her, but Marilyn unknowingly keeps on talking. Others in the staff room can see Mary Ellen's upsetting behavior.

    Using what we know about basic needs what can Marilyn, Mary Ellen and the staff do to deal with this situation? Will our staff agreement be helpful? What can Mary Ellen do to get her needs met? How will Marilyn and the other staff members get their needs met?

## MY PLAN FOR APPLYING THE CONCEPTS . . . Staff

9. If using the new staff agreement for how we plan to treat each other can help everyone meet his or her basic needs in a responsible way, what do I plan to do this week as I work with others that will help me in my relationships at school? What will I do this week to use our new staff agreement? **Decide on a plan (remember SAMS) and record on the Staff Planning and Self-Evaluation page 45, now please.**

## FROM THE STAFF ROOM TO THE CLASSROOM . . .

This week we will revisit our first five plans for meeting our basic needs and agree on a plan for how we are going to treat each other. This is the foundation for establishing our relationships as responsible human beings in our classroom. Taking the time to come to consensus and develop ownership of the agreement will be beneficial throughout the year.

## STAFF: MY PERSONAL PLAN
### For How to Treat Each Other

My plan to apply the concepts personally:

What do I **plan to do** this week as I work with others that will help me in my relationships at school? What will I do this week to apply our new staff agreement? **Record plan here:**

---

## STAFF SELF-EVALUATIONS AND REFLECTIONS . . .

◼ How well did I follow my plan for how I want to treat others? What do I plan to do next?

◼ On a scale of 1 to 10 where am I in getting **all of my needs** met at school?

1 ---------------------------------- 5 ---------------------------------- 10

◼ How did the discussion about Dr. Glasser's one rule, **"Be courteous,"** go in the classroom this week? What did my class agree on for their plan? Write it here:

---

## SUCCESSES On My Journey to Quality I Want to Share With Others:

# STAFF OBSERVATIONS 6
## Agreeing How to Treat Each Other

1. On my journey to quality what did I do that worked well this week in my relationships with others at school?

2. What would I do differently next time?

3. What is my biggest concern? What help do I need?

## Ideas I Developed for How We Treat Each Other . . .

**Bring your journal to share at your next Journey To Quality staff discussion.**

# THIS WEEK IN THE CLASSROOM

## OUTCOMES: Agreeing How to Treat Each Other
- to review the five basic needs
- to combine our class plans for meeting needs into one agreement for responsible behavior
- to introduce Dr. William Glasser, author of <u>The Quality School</u>, <u>Control Theory</u>, and <u>Control Theory in the Classroom</u>

## GETTING STARTED WITH STUDENTS

1. Think about a time when you are the happiest. How do you feel? What are you thinking and doing? Does it meet more than one basic need? **Think first by yourself, then share in class discussion.**

2. Ask students to suggest ways they are meeting their individual basic needs at school. **Divide a piece of chart paper, chalkboard or overhead transparency into fourths. Label each box with one of the four psychological needs of belonging, fun, freedom and power. Continue brainstorming and listing.**

3. Would we be happier if people were more courteous to each other in our classroom? Is it possible to meet our individual basic needs and yet be courteous to each other?

**Here's a way you might want to introduce Dr. Glasser to your students:**

I want to tell you about a man who has an idea about how we can agree to treat each other that is easy to remember because it only takes two words. His name is Dr. William Glasser. He is the man who writes books about our five basic needs and responsibility that we've been discussing in our classroom. "**Be courteous**" is Dr. Glasser's suggestion for how we treat each other. Would this suggestion work in our classroom and be easier to remember than all the Basic Need Plans we made? Let's take each of our plans and compare them to Dr. Glasser's "Be courteous" idea. Could the two words "Be courteous" take the place of all our plans for meeting our basic needs and still help us to get what we want? What are your ideas? **Form Cooperative Learning groups to discuss each plan. Come to class** consensus. This discussion and process activity may take several days to complete. Please don't rush this concept.

✎ Record on Chart: **Our Class Agrees . . .**

**Celebrate the Class Agreement by creating posters illustrating it.**

4. Let's think back to the quality classroom we talked about earlier. If we always treated others courteously what would our room be like? What would the environment be like? Being courteous is easier when people are courteous to you. How will you act when someone is not courteous to you? Remember, we ALWAYS choose our own behaviors. It is our choice how we act. **Discuss together.**

5. **THINGS TO TRY . . .**

   ▸ Make a collage of pictures from magazines to show ways we meet our basic needs.

   ▸ Create a bulletin board that is divided into the four psychological needs. Invite class members to bring pictures for display, classifying them as posted. Continue to add to the bulletin board by taking classroom photos during the year.

   ▸ If you won $1,000,000 what would you do with the money? **Brainstorm your ideas.** What basic needs are met by the items on your list? **Label your list by basic needs categories.**

♪ *Notes & "Quotes"*

## MY PLAN FOR APPLYING THE CONCEPTS . . . Students

6. Share one way you will be courteous at school today. **Group sharing.**

   What am I going to **do** to follow our new class agreement? What will **I do** differently? What will **I do** if I am not treated courteously by others? **Discuss plans.**

## APPLICATION OF LEARNING: Working It Out

7. Ron is a new student in our class. How will we help him understand our classroom agreement? **Discuss.**

8. How will you display your new agreement in your classroom without losing sight of all the concept development that took place in the previous weeks? **How could you show this visually in your classroom?**

## STUDENT PLANNING AND SELF-EVALUATION

9. What is something I did this week to use our "Be courteous" agreement at school? Think of one time you were courteous when someone was not courteous to you. How did you feel? What did you think? What did you do? What would you do differently next time? How did this meet your basic needs?

   **Elementary** - Younger children could do a series of sequence pictures showing what happened.

   **Older students** can describe the events in their journal.

♪ *Notes & "Quotes"*

# **Q** JOURNEY TO QUALITY

## Journey 7
## Our Vision of a Quality School

### DISCUSSION OUTCOMES AND LEARNINGS

- to create a picture in our Quality World of what we want to become in our classroom and in our school
- to identify the indicators of quality in our classroom and in our school

Dr. Glasser defines quality as something we are willing to work hard for because it is need-satisfying. Each one of us has a "picture album" in our Quality World of how we want to meet our basic needs and what quality is to us. As we begin to develop Quality Schools we will want to help our students understand the concept of quality.

When schools first discuss quality in the classroom Dr. Glasser suggests that they begin by discussing **what is good, valuable, or enjoyable in life** now rather than using the word quality (**Reference Bulletin #2**). Follow this by discussing the basic needs met by what is good, valuable, or enjoyable in life.

The concept of quality takes time to develop with students and may need several class discussions in order to help students understand that quality is something need-satisfying in their lives and provides a feeling of having accomplished something good when completed.

Our vision of a Quality School is a place where everyone wants to be, where everyone's basic needs can be met responsibly and where students are asked to learn information that they perceive will add quality to their lives. Having a clear and compelling vision of what we want to become as a Quality School will lead us to the action (Total Behaviors) to get what we want. Without a strong vision we may end up somewhere else. Our visions will keep us focused on quality.

**RECOMMENDED READINGS:**

Glasser, William
**The Quality School**, Chapter 5
**Bulletin #2, The Quality School Reference Bulletins**
**Control Theory in the Classroom**, Chapter 1

# JOURNEY TO QUALITY

## DISCUSSION 7 WITH STAFF
## Our Vision of a Quality School

**MAKING THE CONNECTIONS TOGETHER**
**... from Concepts to Practices**

1. Think about everything in your life that is quality. What are the pictures of quality you have in your Quality World? **Make a list.**

   Look back at your list and identify what basic needs are satisfied by the things you have identified as quality. **Share one thing of quality and the basic need(s) it meets with someone sitting near you.** How did you know it was quality?

2. Dr. Glasser defines a Quality School as one that is need-satisfying. If **our** school were to become a Quality School what would we want it to be and what would it look like? **Process first by yourself, listing the indicators of what you would want. Meet in small groups and agree upon the most important indicators for a Quality School. Share in the large group and chart.**

   ✎ Record on Chart: **Quality School - What We Want**

3. Look at the characteristics we want in a Quality School. What is quality in our school now? On a scale of 1 - 10, how much quality do we have now? **Discuss together.**

   What are some things we can do to get more quality in our school? **Discuss and record.**

   ✎ Record on Chart: **Ways to Get More Quality**

4. What is one thing we can plan to do this week as staff members to move toward more quality in our school? A plan will be successful if it is **s**imple, **a**ttainable, **m**easurable and **s**pecific (SAMS). **Discuss and chart.**

   ✎ Record on Chart: **Our Staff Plan for More Quality**

5. How will our staff agreement for how we will treat each other help us move toward quality this week? **Discuss.** (Refer to last week's discussion of how we will treat each other and Dr. Glasser's one rule of courtesy that can be used in the classroom.)

## WORKING IT OUT WITH STAFF

6. Whenever the staff discusses new ideas and what they could plan to do to improve and move toward quality, Betty says, "Yes, but it won't work. I've tried that before and . . ." This immediately stops the brainstorming of ideas from the group.

What can the principal and staff do to find out more about what Betty wants? What are the things Betty cares about? What need is she meeting by identifying reasons something won't work? How can the staff influence her to be open to new ideas? **Discuss and share ways to influence others.**

## MY PLAN FOR APPLYING THE CONCEPTS
### . . . Staff

7. What is something I **plan** to do this week to add more quality in my classroom? **Record on Staff Planning and Self-Evaluation page 53, now please.**

## FROM THE STAFF ROOM TO THE CLASSROOM . . .

Making the connection between basic needs and quality is an important concept because both give us information about what we are willing to work hard for in our school. In school, students must be able to see the usefulness of the work they are being asked to do and how it will apply to their lives. If they believe it will add quality to their lives (is need-fulfilling) they will be more likely to do it. The same applies to all of us. We will work hard for those things that satisfy one or more of our basic needs and we perceive will add quality to our lives.

♪ *Notes & "Quotes"*

## STAFF: MY PERSONAL PLAN
## For Our Vision of a Quality School

My plan to apply the concepts personally:

What is one thing I plan to do this week to add more quality in my classroom? **Record plan here:**

# STAFF SELF-EVALUATIONS AND REFLECTIONS . . .

◘ How well did I follow my plan for creating quality in my classroom this week? Was my plan **s**imple, **a**ttainable, **m**easurable and **s**pecific? What will I do next week?

◘ Was I able to begin to develop the concept of quality with the students in my classroom this week? What was surprising about the discussions of quality? Did the activities and discussions help me meet the outcomes for the week?

◘ What words did my students use to describe quality?

## SUCCESSES On My Journey to Quality I Want to Share With Others:

# STAFF OBSERVATIONS 7   Our Vision of A Quality School

1. On my journey to quality what did I do that worked well this week?

2. What would I do differently next time?

3. What is my biggest concern?  What help do I need?  What can I do?

## Ideas I Developed to Move Toward Quality . . .

**Bring your journal to share at our next Journey To Quality staff discussion.**

# THIS WEEK IN THE CLASSROOM

## OUTCOMES: Our Vision of A Quality School

- to create a picture in our Quality World of what we want to become in our classroom and in our school
- to identify the indicators of quality in our classroom and in our school

## GETTING STARTED WITH STUDENTS

1. Think about a day in our school and everything that is **enjoyable** for you. **Make a list of what is enjoyable.** The teacher may record on a chart, if preferred. Why is this enjoyable to you? What basic needs did you meet? **Discuss in large group.**

2. Have the students form Cooperative Learning groups and give each group several magazines to share. Ask students to select something they agree is **very good, valuable and enjoyable and that they would really like to have.** Each group shares what it selected with the class and tells why it made that selection. **The teacher records the words students use to describe why something is good, valuable and enjoyable to them.**

   ✎ Record on Chart: **Words That Mean Good, Valuable and Enjoyable**

3. **Display the list of student words.** During the week when students are doing their regular school work or participating in any school activity, ask them to compare what they are doing to the list of words on the chart. Students will be asking, **"Is this enjoyable, valuable or good for me?"** (In other words, is this need-satisfying?) Providing opportunities for short frequent comparisons during the week will help students begin to develop the concept of what is quality for them at school.

4. Later in the week after students begin to show an understanding of the concept of quality, ask them to identify what they would want to see people **doing** in a quality classroom. **Discuss and chart.** (Refer to last week's Class Agreement on how we will treat each other, which is a beginning for attaining quality in our classroom.)

*CHAPTER 7  Page 55*

✎ Record on Chart: **Actions in a Quality Classroom**

5. If we have a picture in our minds of what people would be doing in a classroom working toward quality, what can we **plan** to do this week (from our chart) that will help us move toward quality? **Discuss and decide together on a plan for doing one thing this week that will move the class toward quality. Chart and post.**

✎ Record on Chart: **Quality: Our Class Plan**

**Keep charts up for reference. Continue to make connections between classroom behaviors and indicators of quality throughout the following weeks.**

## MY PLAN FOR APPLYING THE CONCEPTS . . . Students

6. Remind students that we've made a class plan for working toward quality. This is a time to think about specific actions, as individuals, that will help our class effort. What can I plan to do that will be enjoyable, valuable or good for me this week at school? **Share plans in class discussion, remembering SAMS.**

## APPLICATION OF LEARNING: Working It Out

7. Gary says he is bored at school. He never seems to like anything that we do. Based upon what we know about basic needs, what is Gary really saying? **Discuss. Create other scenarios to work out together.**

## STUDENT PLANNING AND SELF-EVALUATION

8. **Continue to invite students to share self-evaluations of their individual plans in class meetings or discussions. They may also choose to reflect privately in a journal or discuss with a friend.**

What is the most enjoyable, valuable or good thing I did this week at school? What basic needs did it meet for me?

- How well did my plan work for me?
- How will doing this add quality to my life?
- What will I be willing to do next week at school that will add quality?

# JOURNEY TO QUALITY

## Journey 8
## Quality Assignments and Quality Work

### DISCUSSION OUTCOMES AND LEARNINGS
- to identify the attributes of quality work
- to identify the attributes of a quality assignment

Dr. Glasser defines quality education as when "what you are learning increases your ability to satisfy one or more of your basic needs." **(Reference Bulletin #11)** The journey to quality education will take time for students to see the value of what they are being asked to do. It will also take time for teachers to create assignments that are seen as need-fulfilling and have value to the student.

Setting expectations for doing quality work is one of the foundational principles of creating a Quality School. Students may resist working hard enough to produce quality work when the work they are asked to do is not seen as need-satisfying to them. Students may be more likely to do quality work when they can make the connection between the assignment and its value to them now or later in life. **(Reference Bulletin #3)** Moving to quality is a process and processes take time.

The **responsibility** for developing quality assignments rests with the teacher and the responsibility for doing quality work belongs to the student. Quality is everyone's job in school.

RECOMMENDED READINGS:

Glasser, William
**The Quality School**, Chapter 7
**Bulletins #3 and #11, The Quality School Reference Bulletins**
**Control Theory in the Classroom**, Chapter 7

# JOURNEY TO QUALITY

## DISCUSSION 8 WITH STAFF
## Quality Assignment and Quality Work

### MAKING THE CONNECTIONS TOGETHER
### . . . from Concepts to Practices

1. Think about yourself as a learner. When and how do you learn best? Take a few moments alone to think about your personal style. **Share how you learn best with the large group.**

2. As a learner you probably know that it's easier to learn when you have a reason and you can see its relevance to your life. The same is true with students. If we want students to do quality work then we must provide quality assignments for them to do. What does a quality assignment look like? **Work first in a small group. Identify attributes of a quality assignment. Share in a large group and chart. Jot down the most significant attributes you want to remember.**

    ✎ Record on Chart: **Attributes of Quality Assignments**

3. Quality assignments are need-satisfying for students because what we are asking them to learn is useful to them. **Label what needs are being met by these attributes on the chart.**

4. If we know what attributes are present in a quality assignment, what would we **see** and **not see** in a quality assignment? **Discuss in a large group. Record on chart and display during the week.**

    ✎ Record on Chart: **Quality Assignments:**
    **We Would See   We Wouldn't See**

5. What can we as a staff plan to do to create quality assignments in our classrooms? **Discuss and agree on what we will do this week that will help us move toward quality assignments.** Post the plan in the staff room.

    ✎ Record on Chart: **Our Plan for Quality Assignments**

Record staff plan for quality assignments here:

## WORKING IT OUT WITH STAFF

6.  Sherry is a new teacher in the building who missed the staff discussion on quality work and quality assignments. She comes to the weekly team planning meeting excited to share the book of black-line masters she bought over the weekend. "There are lots of good pages we can use to keep students busy when they get their work done or we could use them for extra homework," she explains to her team.

How can Sherry's team help Sherry meet her needs for power, belonging, and freedom and yet not compromise their ideals about quality assignments and quality work for students?. **Discuss.**

## MY PLAN FOR APPLYING THE CONCEPTS . . . Staff

7.  What is one thing I plan to do this week to add quality to the assignments I ask my students to do? Is there one thing I plan to do differently? **Record on Staff Planning and Self-Evaluation page 60, now please.**

## FROM THE STAFF ROOM TO THE CLASSROOM . . .

The more you can put yourself in the role of the learner, the more empathy you'll have for students as they struggle to see the usefulness of what they're asked to learn. **This connection between learning and its use is the foundation for students working toward quality.** As the teacher who makes the assignment you'll be rethinking or evaluating it in terms of usefulness as you plan. You will find it easier to communicate this to your students when you've thought it out during the preparation stage. **You might ask your students to think of ways they believe the assignment will be helpful or useful to them.**

## STAFF: MY PERSONAL PLAN
### For Quality Assignment and Quality Work

My plan to apply the concepts personally:

What is one thing I plan to do this week to add quality to the assignments I ask my students to do? Is there one thing I plan to do differently? **Record plan here:**

---

## STAFF SELF-EVALUATIONS AND REFLECTIONS . . .

◘ How well did I follow my plan to add quality to my assignments this week? How did I know they were need-fulfilling for my students?

◘ What did I change, add or delete to my assignments this week to get more quality? Was I able to make the connection of the assignment and its usefulness to the student?

◘ How did my students define quality work? Please include all of their ideas.

◘ If I could make an improvement on what I did this week, what would I do?

---

## SUCCESSES On My Journey to Quality I Want to Share With Others:

# STAFF OBSERVATIONS 8
## Quality Assignments and Quality Work

1. On my journey to quality what did I do that worked well this week?

2. What would I do differently next time?

3. What is my biggest concern? What help do I need? What can I do?

## Ideas I Developed for Quality Assignments and Quality Work

**Bring your journal to share at our next Journey To Quality staff discussion.**

# THIS WEEK IN THE CLASSROOM

## OUTCOMES Quality Assignments and Quality Work
- to identify the attributes of quality work
- to identify the attributes of a quality assignment

## GETTING STARTED WITH STUDENTS

1. Think about your school work. Sometimes you are willing to work hard on an assignment and other times you may not choose to do your best work. What's the difference? **Discuss with class.**

   ✎ Record on Chart: **Assignments When I Work Hard**
   **Assignments When I Don't Work Hard**

2. Dr. Glasser believes that the reason students work hard is because the work meets one or more of their basic needs. Let's look at the assignments identified on the chart when students choose to work hard and see if we can identify what basic needs they might be meeting. **Code the chart, identifying the basic needs.**

3. ♪ **NOTE TO THE TEACHER:** Select an assignment from the chart (Assignments When I Work Hard) that could help you make a clear connection for the students between the assigned work and its value to them now and/or in the future. **Ask students how doing this work will be useful to them now and/or in the future.** Suggested assignments might include spelling which will help in writing letters, or addition and subtraction will help them in balancing a checkbook, etc.

4. Dr. Glasser suggests that it is the **teacher's responsibility to make this connection** between students' work and its usefulness to them in later life, therefore showing the work has value. **In your future lesson planning continue to look for ways to help students make these connections.**

5. If we are willing to work hard on assignments that meet our basic needs, what does the work we do look like? If you were the teacher in our classroom how would you describe quality work to the students? **Brainstorm what it would look like.**

*CHAPTER 8   Page 62*

✎ Record on Chart: **Quality Work**

6. **Work toward developing a class definition of quality from the list that describes quality work. Post in room.**

7. Follow up the discussion on quality work by discussing what we **would** and **wouldn't** see if students were doing quality work in our classroom. **Discuss, encouraging specificity. Chart.**

    ✎ Record on Chart: **We Would See . . .   We Wouldn't See. . .**

## MY PLAN FOR APPLYING THE CONCEPTS . . . Students

8. Think of the assignment you enjoyed doing most today and the one you were willing to work hardest on. Why were you willing to work hard on this assignment? **Discuss in large group.**

## APPLICATION OF LEARNING: Working It Out

9. When its time for spelling (or any subject you choose), Mary always groans and says, "Oh, no! Not that again!" What is she thinking? What doesn't she know? What are different choices she can make when its time for spelling? What are ways to help Mary see the usefulness of spelling and its value in her life?

♫ *Notes & "Quotes"*

# STUDENT PLANNING AND SELF-EVALUATION

10. Ask students to select one example of their work this week that comes the closest to matching the class definition of quality. Have students design a folder for saving their quality work at school (quality keeper). This can be an excellent starting point for self-evaluation. Students could bring the work they identify as quality to share with other students in a class meeting/discussion or one-on-one with a classmate. Some questions you might include are:

   ▸ How will what I learned while doing this assignment be useful to me?
   ▸ How did I decide this was quality work for me?
   ▸ Why did I choose to work hard on this assignment?
   ▸ It is harder to do quality work?

♪ *Notes & "Quotes"*

# JOURNEY TO QUALITY

## Journey 9
## The Expanded Role of the Teacher

**DISCUSSION OUTCOMES AND LEARNINGS**

- to understand that teachers relate to students in four ways: as friend, counselor, teacher, manager
- to understand Dr. Glasser's definition of teaching
- to expand the understanding of the role of friendship in the classroom

Our relationship with students is broader than teaching. There are four roles, as defined by Dr. Glasser, in any relationship: **friend, counselor, teacher** and **manager**. Rather than viewing these roles as getting in the way of teaching, we can envision a broader concept of it. It is through these overlapping relationships that **teachers can help students see that learning can add quality to their lives.** If a student views a teacher as a friend, the teacher may be able to influence the student to make different choices.

**Friendship is the most basic relationship** because everyone has a genetic need to belong. We become friends by finding a common ground to share and act upon. Student learning is the basis for friendship in teaching. It is the common ground for building relationships at school.

Counseling and friendship go hand in hand. We cannot counsel unless we have first established a friendship. The counseling role provides teachers with a communication strategy (Reality Therapy) for helping students **develop responsible behavior.** This is accomplished by focusing on the students' present behaviors, choices they can make and ways they can evaluate their own behaviors.

**The role of the teacher is teaching.** It is defined by Dr. Glasser as "the process of imparting specific skills and knowledge through a variety of techniques, like explaining and modeling, to people who want to learn these skills and knowledge because they believe that, sooner or later, these skills and knowledge will add quality to their lives." **(Reference Bulletin #11)**

Through friendship, teachers get insights into students' values and can help them make the connection between what is being learned and the value it will have in their lives. One reason teaching is difficult is because students may not have perceived the value of education.

The teacher-as-manager role is to persuade students to do quality work by giving them information. Managing has been difficult because some teachers have used a boss-manager, external control, approach to get students to do school work. Those using stimulus response believe we can make students do what we want them to do by using external control. If we want students to do quality work we will need to manage students differently. We will need to become lead-managers using persuasion, information and helping students make different choices.

**RECOMMENDED READINGS:**

Glasser, William
**The Quality School, Chapter 11**
**Bulletins #1 and #11, The Quality School Reference Bulletins**

# JOURNEY TO QUALITY

## DISCUSSION 9 WITH STAFF
## The Expanded Role of the Teacher

### MAKING THE CONNECTIONS TOGETHER
### ...from Concepts to Practices

1. Think about your role as a teacher. What are you doing now that you **didn't** think was a part of teaching when you first became a teacher? **Share in a small group, then discuss as a staff and chart.**

   ✎ Record on Chart: **More Than Teaching**

2. From this list on the More Than Teaching chart, **code the responses** according to the four roles Dr. Glasser identifies in any relationship (**f**riend, **c**ounselor, **t**eacher, **ma**nager.)

3. What implications does this have for teachers? Are we doing more of one role than the others? **Discuss.**

4. Think back to the **Journey To Quality** we've been taking and what we've been doing in our classrooms to establish a friendly environment. If this is paramount how are we continuing to **intentionally** establish our role of friendship with students? What things could we plan to do now and in the future that will help us continue to develop involvement and friendship with students in our school and classrooms? **Discuss and chart ideas.**

   ✎ Record on Chart: **Ideas for Involvement**

5. From our list of ideas what will we as a staff plan to do this week that will increase involvement and friendship with students? **Discuss options and agree on a plan together,** remembering that a successful plan is **s**imple, **a**ttainable, **m**easurable and **s**pecific (SAMS). **Chart and post.**

   ✎ Record on Chart: **Our Plan For Involvement**

6. Record the Staff Plan for Involvement here:

## WORKING IT OUT WITH STAFF

7. Terry is a substitute teacher who has a "school bag" full of ideas for any class she's called to teach. Most often she doesn't have time to get any of them out to use because the regular teacher has left so much "busywork" for the students to complete.

   Terry feels she has to choose between following the teacher's plans or spending some learning time on "fun" activities that will help her relate as a friend with the class. How could Terry "work it out" in her role as a substitute teacher? **Discuss.**

## MY PLAN FOR APPLYING THE CONCEPTS . . . Staff

8. What is one thing I plan to do this week to help increase my involvement and friendship with the students in my classroom? **Record on Staff Planning and Self-Evaluation, page 69, now please.**

## FROM THE STAFF ROOM TO THE CLASSROOM . . .

How we relate to each other is the basis of friendship and involvement in our classroom, not only for student-to-student, but teacher-to-student. Establishing friendship is the prerequisite for developing the other roles of a relationship: counseling, teaching and managing. This week we will expand the role of friendship with students in the classroom. Friendship is an important way of meeting the genetic need everyone has for love and belonging. Helping students to learn the importance of friendship and ways to build friendship and involvement is useful for meeting their need for belonging. Learning skills of cooperation can build better relationships in the classroom which ultimately could lead to developing friendships.

*♪ Notes & "Quotes"*

# STAFF: MY PERSONAL PLAN
## For The Expanded Role of the Teacher

My plan to apply the concepts personally:

What is one thing I plan to do this week to help increase my involvement and friendship with my students? **Record plan here:**

## STAFF SELF-EVALUATIONS AND REFLECTIONS . . .

◘ How well did I follow my plan for increasing my involvement as a friend with my students this week? What will I do next?

◘ What insights about my students did I gain from this week's focus on involvement and friendship?

◘ Now that I've had an opportunity to think about the expanded roles of a teacher, what am I **really** thinking about these roles as . . .

▸ Friend:

▸ Counselor:

▸ Manager:

▸ Teacher

## SUCCESSES On My Journey to Quality I Want to Share With Others:

# STAFF OBSERVATIONS 9
## The Expanded Role of the Teacher

1. On my journey to quality what did I do that worked well this week?

2. What would I do differently next time?

3. What is my biggest concern? What help do I need? What will I do?

## Ideas I Developed for Involvement and Friendship . . .

**Bring your journal to share at our next Journey To Quality staff discussion.**

# THIS WEEK IN THE CLASSROOM

## OUTCOMES  The Expanded Role of the Teacher

- to understand that teachers relate to students in four ways: as friend, counselor, teacher, manager
- to understand Dr. Glasser's definition of teaching
- to expand the understanding of the role of friendship in the classroom

## GETTING STARTED WITH STUDENTS

1. Think back over all the friendships you have had. What makes a really good friend? Write down what you like in a best friend. **Discuss in large group. Record attributes of a friend on a chart.**

    ✎ Record on Chart:  **A Friend . . .**

2. **Continue to work on the chart of attributes of a good friend through the process of brainstorming, grouping and labeling. Continue to refine the list until you have distilled friendship into three or four key points.**

3. Use Cooperative Learning groups this week so that students can work with a variety of classmates. This opportunity to work together in new groups can serve as a basis for starting to develop new friendships.

4. Provide writing assignments where students write about their friends. Add to these examples:

    ▸ What to do to have a best friend
    ▸ What to do if a friend moves away
    ▸ What to do if a friend doesn't like you anymore
    ▸ What to do if a friend starts doing a new activity that doesn't interest you

5. **Ask students to discuss involvement and why we enjoy having the opportunity to**

CHAPTER 9    Page 71

**work with our friends.** Given what we know about friendship, what does this tell us about what we would need to do to have more friends? **Have the students share their ideas about ways to get more friends. Chart.**

✏ Record on Chart: **Ways To Get More Friends**

## MY PLAN FOR APPLYING THE CONCEPTS . . . Students

6. Think about all that we have talked about with friendship. What am I willing to plan to do this week that will help build involvement and friendship with people in my class? **Share your plan with the class or write about it in your journal. Younger students could draw pictures or a series of pictures showing what they enjoy doing with their friend(s).**

## APPLICATION OF LEARNING: Working It Out

7. Melanie wants Clare to be her friend, but Clare doesn't want to be friends with Melanie. How can Clare handle the situation without hurting her feelings? What can she do?
   ♪ **NOTE TO THE TEACHER: Adjust the situation to meet the age and needs of the students that you teach.**

## STUDENT PLANNING AND SELF-EVALUATION

8. During this week think about your friendships and ask yourself where you are on this friendship scale. **Put an X on the scale where you place yourself now.**

FRIENDSHIP SCALE

| Want **More** Friends | Have **Enough** Friends | Have **All** I Want |
|---|---|---|
| 1    2    3    4 | 5    6    7 | 8    9    10 |

9. Are you satisfied with where you are on the Friendship Scale? If not, what can you do? **Discuss some choices you could make. What will you plan to do now?**

**Q** **JOURNEY TO QUALITY**

## Journey 10
## Quality Manager: The Role of the Teacher

**DISCUSSION OUTCOMES AND LEARNINGS**

- to understand Dr. Glasser's definition of the teacher as a manager
- to compare external control with internal locus of control

When we began teaching, most of us believed our only role was to teach, but we soon found much of our time was being spent in doing other things. Dr. Glasser identifies four ways we relate to each other that are evident in any relationship. The four roles are teacher, manger, counselor and friend. When we understand how these roles and relationships function we can gain insight into how we can have more time for teaching.

The role of the teacher as a manager is defined by Dr. Glasser as "the process of convincing people that working hard and doing a quality job of what the manager (teacher in the case of school) asks them to do will add quality to their lives, and usually, to the lives of others." **(Reference Bulletin #11)**

Managing is considerably different from teaching and more difficult. Dr. Glasser states in **The Quality School** that **teaching is the most difficult profession** because teaching is one of the few professions that has to deal with resistors. In other professions, such as medicine or hair styling, people use the services by choice. But in school, students are usually **not** there by choice. Often students do not perceive that what they are asked to do meets their basic needs. They also may not see that the work they are asked to do will add quality to their lives or be useful to them.

In schools we used to believe that we could control student behavior using a stimulus-response approach. Dr. Glasser's Control Theory is the opposite view. **Control Theory is a theory of human behavior which is based upon the belief that each of us is internally motivated, and we choose our behaviors to meet our basic needs. We cannot externally control others.**

The role of the teacher-as-manager is to help "students see a strong connection between what they are asked to do and what they believe is quality."

Dr. Glasser defines coercion as **anything** we do to try and control another person's behavior. If we are unsure if what we are doing is coercive, we need to ask ourselves why are we doing it and, if we are trying to control the student behavior. Many of us have operated in schools for years using external control systems to manage student behavior. Rewards, stickers, and other methods of reinforcements often have been used to **attempt** to control behavior. Being aware of our reasons for our own behavior may be a first step in eliminating coercion in the classroom.

Eliminating coercion in the classroom is difficult because it requires a change in our thinking and our action. An understanding of Control Theory/Reality Therapy and how it applies to our lives is a good place to begin. As we gain a new perspective we can begin to evaluate some of our classroom practices that may be coercive.

Dr. Glasser says to keep this statement in mind: "Coercion does not increase the quality of anyone's life. Coercion doesn't work because it always reduces the quality of the lives of the students who are coerced." In other words, coercion reduces the ability of students to meet their basic needs in satisfying ways. When students are coerced they may attempt to meet their basic need for power in a coercive way. This can get in the way of others getting their basic needs met.

In a Quality School, the role of the teacher shifts from a boss-manager, top down, stimulus-response approach of managing to a lead-manager, non-coercive approach. We recognize from Control Theory that we cannot make or force anyone to do anything. The lead-manager's role is first to establish a relationship of friendship with students in the classroom based on the common ground of learning. The other roles of counseling, teaching and managing are also built upon friendship. In a Quality Classroom and a Quality School the teacher as an effective lead-manager persuades the students non-coercively that what they are being asked to do is important, need-fulfilling and will add quality to their lives.

**RECOMMENDED READINGS:**

Glasser, William
**The Quality School**, Chapters 2 & 11
**Bulletin #11, The Quality School Reference Bulletins**
**Control Theory in the Classroom**, Chapter 7

# JOURNEY TO QUALITY

## DISCUSSION 10 WITH STAFF
## Quality Manager: The Role of the Teacher

### MAKING THE CONNECTIONS TOGETHER
### ...from Concepts to Practices

1. Think about a teacher that you were willing to work hard for when you were a student. As you remember this teacher, analyze the reasons that contributed to your desire to work hard. **Make a list of reasons, first for yourself, and then share with the group. Chart the responses.**

    ✎ Record on Chart: **Why I Worked Hard**

2. Look back at the "Why I Worked Hard" chart. What can you determine were reasons for your hard work? Think in terms of Control Theory. **Record reasons** on the chart using a different color pen. Example: The teacher liked me = Belonging Need.

3. Now think about a teacher who did not create a desire to work hard when you were a student. Examine your reasons and chart and label in RT/CT terms. Example: didn't respect the teacher = not in the student's Quality World.

    ✎ Record on Chart: **Why I Didn't Work Hard**

4. Using Dr. Glasser's definition of managing, think again about the two teachers you remember when you were a student and see how this definition applies:

    **MANAGING - persuading people that working hard and doing a quality job of what the manager (teacher in case of the school) asks them to do will add quality to their lives and, usually, to the lives of others.**

What did you discover about why you were willing to work hard for one teacher and not the other? What did you discover about coercive practices? **Share with the group.**

5. Are there some practices with students in our school or classrooms that could be viewed as coercive? **Identify practices and record on chart.**

    ✎ Record on Chart: **Coercive Practices Toward Students**

6. Is coercion okay to use toward adults? Do we have coercive practices in our school toward staff? **Identify, discuss and record on chart.**

    ✎ Record on Chart: **Coercive Practices Toward Staff**

7. If our goal is to manage without coercion what will we plan to do? **Discuss and agree on a plan we can do this week that will lead toward eliminating coercion in our school.** (Remember SAMS: simple, attainable, measurable, specific.)

    ✎ Record on Chart: **Our Plan to Eliminate Coercion**

8. Record the Staff Plan for beginning to eliminate coercion here:

## WORKING IT OUT WITH STAFF

9. Jennifer has read all of Dr. Glasser's books and believes that eliminating coercion and fear is right. She continually evaluates her own behavior and is doing well on her personal plan to eliminate coercion, yet in her classroom she is having difficulty giving up the external controls for managing student behaviors. Why does Jennifer still feel the need for coercive practices within the classroom? What help does she need to make the translation from personal application to the classroom application?

## MY PLAN FOR APPLYING THE CONCEPTS ... Staff

10. Make a list of practices that may be coercive in your classroom. Look at your list and decide how you can begin to eliminate coercion. Think of **one thing** you plan to do differently this week. **Record on Staff Planning and Self-Evaluation page 78, now please.**

# FROM THE STAFF ROOM TO THE CLASSROOM . . .

We may have students in our classroom who are very resistive. We often refer to these students as the "behind-the-line-kids." This implies that these students are not in a workable area where we can persuade them to make better choices. It is through friendship that we can begin to move these students into an area where we can influence them. Friendship is the basis for all relationships, and it will create a starting point for the student and teacher to establish a working relationship.

# STAFF: MY PERSONAL PLAN
## Quality Manager: The Role of the Teacher

My plan to apply the concepts personally:

- Practices that may be coercive in my classroom are: **(List here)**

- One thing I **plan** to do differently this week to eliminate coercion is: **(Record plan here)**

# STAFF SELF-EVALUATIONS AND REFLECTIONS . . .

- How well did I follow my plan to eliminate coercive practices in my classroom?

- What have I learned this week about my attempts to control student behavior? What surprised me?

- What changes am I considering? What do I plan to do next?

# SUCCESSES On My Journey to Quality I Want to Share With Others:

# STAFF OBSERVATIONS 10
## Quality Manager:  The Role of the Teacher

1. On my journey to quality what did I do that worked well this week?

2. What would I do differently next time?

3. What is my biggest concern?  What help do I need?  What can I do?

## Ideas I Developed About Managing and Non-Coercion . . .

**Bring your journal to share at our next Journey To Quality staff discussion.**

# THIS WEEK IN THE CLASSROOM

## OUTCOMES  Quality Manager: The Role of The Teacher
- to understand Dr. Glasser's definition of the teacher as a manager
- to compare external control with internal locus of control

## GETTING STARTED WITH STUDENTS

We are deviating from the usual format of THIS WEEK IN THE CLASSROOM because the concept of managing without coercion is such a pivotal point in developing a Quality School. This week teachers will be focusing on understanding coercion and analyzing their classroom practices as **coercive** or **non-coercive** in their role as manager. It may be helpful to keep Dr. Glasser's definition of coercion in mind: **Coercion is anything we do to control another person's behavior.**

Teachers may also want to revisit any of the key concepts previously taught in **The Journey to Quality** during this week.

# JOURNEY TO QUALITY

## Journey 11
## Coercion: Positive and Negative

---

**DISCUSSION OUTCOMES AND LEARNINGS**

- to understand why all coercion, positive and negative, must be eliminated in a Quality School
- to understand the role of influence and persuasion in managing student behavior
- to apply the Total Behavior System to choices we make

---

If we believe in Stimulus Response Theory (**external** control) we believe that we can control students' behavior by what we do **to** and **for** them. If we believe in Control Theory (**internal control system**) we believe that we cannot control students' behavior. **All we can do is to take more effective control of our own behavior.** We can only influence or persuade students if we are a part of their Quality World.

When students are coerced to do things they do not want to do they become increasingly resistant. Not only do they resist the school work, they become more resistant toward the teacher. Student energies are spent resisting any attempts to control their behaviors.

Positive coercion is externally reinforcing behavior we want to see continue. Negative coercion is when punishment or consequences are used to eliminate behaviors. Both are coercive because they are attempts to control or "shape" student behavior. This stimulus response thinking does not hold students responsible for controlling their own behavior.

According to Control Theory we can manage students more effectively when we are in their Quality World. Friendship is the basic way we can get into a student's Quality World. Once inside a student's Quality World, the teacher/manager can persuade or influence students that the work they are asked to do will be need-fulfilling, add quality to their lives and be useful to them now or later. In Control Theory the way we behave is always our best attempt to meet our basic needs. Our behavior system consists of four behaviors:

- ▶ **Physiology** - the way our body responds
- ▶ **Feelings** - the emotions we experience
- ▶ **Thinking** - the thought process we go through
- ▶ **Doing** - the actions we take

Our behavior system operates much like the wheels on a car. All four wheels/behaviors are operating at the same time. Our behavior system in Control Theory terms is called the Total Behavior System because all of our behaviors always operate together at all times.

**RECOMMENDED READINGS:**

Glasser, William
**The Quality School**, Chapter 3
**Bulletins #1 and #3, The Quality School Reference Bulletins**
**Control Theory in the Classroom**, Chapter 7

# JOURNEY TO QUALITY

## DISCUSSION 11 WITH STAFF
## Coercion: Positive and Negative

### MAKING THE CONNECTIONS TOGETHER
### ...from Concepts to Practices

1. Think back to when you first started teaching and ways you may have tried to control student behaviors. Were there ever times when using coercive (stimulus response) practices seemed to get you what you wanted? **Discuss with a person sitting next to you; then share with large group.**

2. Stimulus Response Theory is based upon the belief that we can control another person's behavior using an external stimulus. What about stickers, tokens and class parties when the classroom earns a certain number of points? Are these **positive stimulus response** activities coercive? Is positive coercion okay? **Discuss.**

3. Dr. Glasser believes that **any** attempt to control another person's behavior is coercive and that **any coercion, positive or negative**, denies that person the opportunity to take responsibility for his/her own actions. **Control Theory is based upon the belief that we are all internally controlled and thus responsible for our own choices.**

When a student is not using responsible behavior, we always have a choice on how we act or behave toward the student.

When you observe a student not following the class agreement, do you ever experience a physiological signal or urge to behave? Which wheels are you aware of on your Total Behaviors Car? What do you need to do to make wise choices in your behaviors when you are aware of the signal? **Discuss with the large group.**

4. What approaches are available for us in school when we accept Control Theory? How can we manage student behavior successfully if we eliminate coercion? **Discuss and record.**

   ✎ Record on Chart: **Strategies for Managing Non-Coercively**

5. Is managing non-coercively something I am willing to work hard for in order to help students take responsibility for their own behavior? **Discuss with someone sitting near you.**

6. What can we as a staff **plan** to do to begin replacing coercive strategies with non-coercive ways of managing students? **Look at the staff list of ideas on Strategies for Managing Non-Coercively. Discuss and agree on what we will do this week together that will help us move toward non-coercion and quality.** Post the plan in the staff room.

    ✎ Record on Chart: **Our Plan for Non-Coercive Managing**

7. Record the Staff Plan for Non-Coercive Managing here:


## WORKING IT OUT WITH STAFF

8. Kathy always assigns a reasonable amount of homework for her students. She provides time for independent practice, helps keep parents informed and follows her school district's policy. Kathy put away her homework stickers when she decided she was really using them to coerce her students into doing the homework. She isn't comfortable with her decision because she doesn't have a plan for how she is going to manage homework without rewards. What if the students don't do their homework now?

What could she do to **begin** managing homework using a non-coercive strategy? What might be a plan she could use that would reduce her anxiety over her decision? How can she shift the responsibility for completing homework to her students? **Discuss.**

**Consider your homework practices:**
1. What is the purpose of homework in my classroom?
2. How can I help students see the usefulness of homework?
3. How can I influence or persuade students that what I'm asking them to do will add quality to their lives?


## MY PLAN FOR APPLYING THE CONCEPTS
### . . . Staff

9. This week as you work with your students, ask yourself: Am I trying to **control** their behavior or am I trying to **influence** them? Am I taking away my students' opportunities for developing personal responsibility by the management strategies I use in the classroom now? What is one thing I **plan to do** this week that will shift responsibility for behavior to my students without reducing my effectiveness as a teacher? **Record on Staff Planning and Self-Evaluation page 86, now please.**

# FROM THE STAFF ROOM TO THE CLASSROOM . . .

A class in **Reality Therapy/Control Theory** can provide an in-depth application of skills and processes for managing students without coercion.

These concepts are not simple and Reality Therapy/Control Theory offers us no "quick fix." It requires a paradigm shift in our thinking away from how we were taught, as well as from the way most of the world believes and acts now. Be gentle with yourself as you begin to apply these concepts in the classroom. It takes time and effort to eliminate coercion.

With a partner spend some time discussing these questions:

- What is wrong with positive and negative coercion?
- Why does positive coercion seem to work?
- Is a **little** coercion okay? Elaborate.

These informal discussions will help to clarify your understanding of the positive and negative coercion concepts.

♫ *Notes & "Quotes"*

## STAFF: MY PERSONAL PLAN
### Coercion: Positive and Negative

My plan to apply the concepts personally:

This week as you work with your students, ask yourself: Am I trying to **control** behavior or am I trying to **influence** behavior? What is one thing I plan to do this week that will shift responsibility to my students without reducing my effectiveness as a teacher? **Record plan here:**

---

## STAFF SELF-EVALUATIONS AND REFLECTIONS . . .

◘ What have I discovered about myself this week about how I manage students?

◘ On a scale of 1 to 10, where am I in eliminating coercion in my classroom?

1 ------------------------------ 5 ------------------------------ 10

| | |
|---|---|
| Stimulus Response | Control Theory |
| Boss Manager | Lead Manager |
| Coercion | Non-Coercion |
| Teacher in Control | Student in Control |
| Power Over . . . | Empowerment |

◘ How well did I follow my plan to shift more responsibility to my students? What will I do next?

---

## SUCCESSES On My Journey to Quality I Want to Share With Others:

## STAFF OBSERVATIONS 11  Coercion: Positive and Negative

1. On my journey to quality what did I do that worked well this week?

2. What would I do differently next time?

3. What is my biggest concern?  What help do I need?  What can I do?

## Ideas I Developed for Eliminating Coercion . . .

**Bring your journal to share at our next Journey To Quality staff discussion.**

*CHAPTER 11*

# THIS WEEK IN THE CLASSROOM

## OUTCOMES   Coercion: Positive and Negative

- to understand why all coercion, positive and negative, must be eliminated in a Quality School
- to understand the role of influence and persuasion in managing student behavior
- to apply the Total Behavior System to choices we make

## GETTING STARTED WITH STUDENTS

1. Think about a time you and your friends got into trouble. Have you ever said, "They made me do it?" **Discuss a time you believe you were talked into doing something. Share in a large group.**

2. Do you believe that your friends can **make** you do things that you do not want to do? **Discuss.**

3. What were some of the other choices you could have made instead of getting into trouble?  ♪ **NOTE TO THE TEACHER: Choose some of the experiences shared in discussion #1 to use as examples. Continue discussing until many choices have been explored. Help the students to see that there are usually a variety of choices available to them.**

4. What really controls our behavior?  **Discuss this question in a class meeting or discussion.**

5. ♪ **NOTE TO THE TEACHER:** Here's a way you might want to talk about behavior and the four parts of our Total Behavior System with your students:

In Control Theory we control our own behavior and our behavior system is made up of four parts. The four parts of our behavior are doing, thinking, feeling, and physiology. All of the four parts of behavior are operating together at all times. Let's look at the four parts as if they are the four wheels on the car. The back two wheels are the feeling wheel and the physiology (body) wheel. The front two wheels are the thinking and doing wheels.

CHAPTER 11    Page 88

Let's imagine that our phone is ringing, and we'll check to see what each wheel could be experiencing:

- **feeling wheel** - excited, happy, scared, etc.
- **physiology (body) wheel** - knot in your stomach, heart beating rapidly
- **thinking wheel** - It's my friend. It's not for me. They will hang up. Maybe it's a wrong number.
- **doing wheel** - Answer it. Don't answer it. Run to the phone. Stay in the bathtub, etc.

Who makes the choices for you when the phone rings? Who makes the choices for you when your friends want you to do something? Do you always have a choice? **Discuss.**

## MY PLAN FOR APPLYING THE CONCEPTS . . . Students

6. Think about something you did today at school and all the other choices you eliminated when you decided to do it. What were you **thinking, feeling** and **doing**? **Discuss as a class together.**

## APPLICATION OF LEARNING: Working It Out

7. Walt is often late for class. He usually arrives with no explanation and creates a disturbance when he enters the classroom.

    1. What is Walt getting out of being late to class? Is it need fulfilling for him?
    2. What could the teacher say to Walt to help him make more responsible choices?

    **Role play the situation or discuss questions the teacher could ask Walt.**

## STUDENT PLANNING AND SELF-EVALUATION

8. This week think about the choices that you made and **who** controlled your behavior. Think about how that was the best choice for you at that time. What were other choices you could have made? Write about or draw a picture showing one or two of them.

# JOURNEY TO QUALITY

## Journey 12
### Choices We Make

**DISCUSSION OUTCOMES AND LEARNINGS**

- to recognize that we can only control ourselves
- to understand that we always have choices for how we behave
- to understand that we can choose to solve problems non-coercively

**Control Theory** is based upon the belief that each of us is an internal control system and that the only person we can control is ourself. Just as the teacher is an internal control system and can only control himself, so is the student. Our old ways of thinking about controlling others will **shift from controlling to influencing**. Helping students gain **self-control and take responsibility** for their own choices (actions) becomes an important role of the teacher.

**Each of us always has a choice about how we behave.** We always act as a Total Behavior System to meet our basic needs. At times it seems as if we are like a front wheel drive automobile with the back wheels of physiology and feelings being the only wheels operating and the front two wheels of thinking and doing not engaged. When we choose behaviors of stressing, headaching, angering, crying and upsetting we are operating on our two back wheels. We are stuck in the sand. It is not until we engage the front two behavior wheels of thinking and doing that we can move forward and make more effective choices.

In any classroom we can be sure problems or incidents will arise. **It is how these problems are handled that is the key to eliminating coercion in the classroom.** Understanding we always have a choice of how we behave is paramount to doing things differently in the classroom.

Agreeing how we will solve problems is important with any group. Dr. Glasser suggests in **Reference Bulletin #14** to develop a class agreement which includes ways to solve problems in the classroom.

Dr. Glasser's statement:
> In our class if anyone, including the teacher, is upset or has any problem, we will talk it over and try to solve the problem. In solving the problem, we will never hurt, punish, criticize or put anyone down. We will always be willing to participate in a whole class discussion and any one can ask for a class discussion at any time. **In other words, we can work out upsets, and problems by talking it over together with courtesy.**

The class may choose to select Dr. Glasser's statement for problem solving or to write its own agreement which will include Dr. Glasser's key points. Perhaps the greatest reason for implementing Control Theory and Reality Therapy in the classroom is to help students develop **responsible behavior**. It is with a caring teacher's understanding of internal locus of control that students can be guided to make responsible choices.

When we take ownership of our own behavior we can see clearly why stimulus response has not worked in the past. In a stimulus response environment the teacher attempts to be the controller of students behavior and provides the consequences, positive or negative, for the behavior.

Conversely, in a Control Theory classroom, the teacher understands that students choose their own behaviors and when problems arise, they can be worked out. Stimulus Response Theory suggests that the teacher can be the solver of problems and Control Theory puts the ownership of behavior and choice with the student. Which approach do you think will develop more responsible students?


RECOMMENDED READINGS:

Glasser, William

**Bulletins #10 and #14, The Quality School Reference Bulletins**
**Control Theory,** Chapter 6
**Control Theory in the Classroom,** Chapter 5

# JOURNEY TO QUALITY

## DISCUSSION 12 WITH STAFF
## Choices We Make

### MAKING THE CONNECTIONS TOGETHER
### ... from Concepts to Practices

1. Think about returning to your classroom after a long break and feeling rested. You say to yourself, "I'm feeling great! I won't get upset today." Yet by 10:00 you begin to feel you want to control the students' behavior. What are some of the choices of behavior you have when you experience this feeling? **Process first by yourself and then in the large group, brainstorm a list of choices you could make. Record on a chart.**

   ✎ Record on Chart: **Our Choices of Behavior**

2. Look at the chart and decide which choices are coercive (mark with a C) and which are non-coercive (mark with an N-C).

3. Look at a typical situation that can happen in any school. You arrive in the morning, and find that an unscheduled staff meeting has been called. This throws your schedule off because you had counted on your planning time to get ready for the day.
   ▸ What are you feeling?
   ▸ What is your physiology?
   ▸ What are you thinking?
   ▸ What do you do? What choices do you have?
   **Discuss with your neighbor.**

4. Review your choices and determine which were effective and non-effective. Whose behavior were you trying to control? **Continue discussing with your neighbor.**

5. Dr. Glasser tells us we always act as a **Total Behavior System,** but sometimes we seem to be operating on our two back wheels of feelings and physiology. This is a time we are more apt to be angering, headaching or complaining, when in fact activating our two front wheels of thinking and doing could lead to more effective choices for getting what we want. What do we know about engaging the thinking and doing wheels in making choices? Consider the following and the choices we could make. **Discuss as a group.**

- You get a new student just as the bell rings.
- A team meeting is called for after school and you have a dentist appointment.
- Your child's school calls and your child is sick and you have no baby sitter.
- The principal asks you to serve on a district-wide committee and you know it will take many meetings outside of school.

6. Dr. Glasser writes there will always be **instances** at school when people have disagreements. He believes a class agreement about how we will solve problems will be helpful. **In Reference Bulletin #14,** Dr. Glasser suggests that a **"work it out"** statement be posted in every classroom. Think about his idea and how this agreement might be useful for the staff. Would an agreement to solve problems without coercion be helpful? **Discuss as a group.**

7. If the staff agrees that solving problems without coercion is important, strive to develop a problem-solving agreement. Focus the discussion on the key concept that **together we can work out problems. Develop an agreement together. Chart and display in staff room.**

✎ Record on Chart: **Work It Out**

8. Could this type of an agreement be taken to the school council, PTSA, other school related groups? Would this be a useful agreement for **everyone** in a Quality School? **Continue discussion with staff.**

## WORKING IT OUT WITH STAFF

**Working It Out focuses on the choices we make this week.**

9. Anita never voices her opinion during a staff meeting, but the moment a decision is made she is up and down the hall talking to her colleagues about why she doesn't think it will work. She often chooses angering and upsetting over issues rather than addressing her concerns during the staff meeting.

    How can Anita move off her "back two wheels" (physiology and feelings) and engage her "front two wheels" (thinking and doing) to help her make better choices? How can staff members be sensitive to Anita's needs and yet help her see she is responsible for her own choices?

♫ *Notes & "Quotes"*

CHAPTER 12     Page 93

## MY PLAN FOR APPLYING THE CONCEPTS . . . Staff

10. If I know that **I can only control myself,** what implications does this have for me as I work to resolve problems within the classroom? What will I plan to do this week to help **my students** be more aware of the choices they make? What will I plan to do this week to help **me** be more aware of the choices I make? **Record on Staff Planning and Self-Evaluation page 95, now please.**

## FROM THE STAFF ROOM TO THE CLASSROOM . . .

We've been talking about the choices we make when we behave and how we are the only person we can control. These key concepts of Control Theory help us better understand our Total Behaviors. It is important to recognize that the choices we make are Total Behaviors and that it is through the engaging of thinking and doing that we are able to make more appropriate behavior choices.

If we want students to make more responsible choices and develop more responsible behavior then their understanding of Total Behaviors is important. Helping students realize that they **always have a choice** in how they behave will be the key point to focus on this week.

♪ *Notes & "Quotes"*

## STAFF: MY PERSONAL PLAN
### For Choices We Make

My plan to apply the concepts personally:

What will I plan to do this week to help **my students** be more aware of the choices they make?  What will I plan to do this week to help **me** be more aware of the choices I make?  **Record plan here:**

---

## STAFF SELF-EVALUATIONS AND REFLECTIONS . . .

◘  Was I able to develop a "work it out" agreement with my students this week? What is our agreement? **Record here.**

◘  Did class participation in developing the problem-solving agreement create a sense of ownership in my class?

◘  Are my students beginning to understand they always have choices in solving problems? What have I observed about my students' understanding of this concept?

◘  How am I doing at identifying the "wheels" of my Total Behavior System? Am I able to make more effective thinking and doing choices in my Total Behaviors?

---

## SUCCESSES On My Journey to Quality I Want to Share With Others:

# STAFF OBSERVATIONS 12  Choices I Make

1. On my journey to quality what did I do that worked well this week?

2. What would I do differently next time?

3. What is my biggest concern?  What help do I need?  What will I do?

## Ideas I Tried for Making Choices . . .

**Bring your journal to share at our next Journey To Quality staff discussion.**

# THIS WEEK IN THE CLASSROOM

### OUTCOMES Choices We Make

- to recognize that we can only control ourselves
- to understand that we always have choices for how we behave
- to understand that we can choose to solve problems with non-coercive behavior

### GETTING STARTED WITH STUDENTS

1. Consider a time that you had a big disagreement with a friend. How did you work it out? Are you still friends? **Share the experience with the group. Teacher records the different ways students solved their disagreements on chart paper or chalkboard.**

   ✎ Record on Chart: **Solving Disagreements**

2. From our list of ways we solved disagreements identify those that followed our class agreement, "Be courteous." **Class discussion. Teacher marks those strategies that are consistent with the class agreement.**

3. What was different between those that were courteous and those that weren't courteous? Which were the more effective ways to solve disagreements? **Discuss together.**

4. We always have a choice for how we solve problems. We can work them out courteously or non-courteously. We know there will always be times when there are problems and incidents. How we behave is our choice.

   ♪ **NOTE TO THE TEACHER:** You might want to use the car analogy as you explain Total Behaviors to your class. Remember the car that we learned about to help us understand how we behave. We have four wheels which are our Total Behaviors. All four wheels are **always** operating. When we have a problem we feel our back wheels the most. The back wheels are physiology (body wheel) and feeling wheels,

CHAPTER 12

and it almost seems like we are doing a "wheelie" when we are angry and upset. Our front two wheels must be engaged before we can solve problems effectively. Our front two wheels are the thinking and doing wheels and they are the key to working problems out.

Can you remember the last time in our classroom when there was a problem and we were operating on our back two wheels and we needed to engage our front two wheels? **Discuss and describe.**

5. In our classroom, just as with our families, there will always be incidents. Would it be helpful if we had an agreement about how we are going to solve problems in our classroom? Dr. Glasser has written a suggestion for people to consider about working things out. This may be helpful to read before we work together to write our own.

> In our classroom if anyone, including the teacher, is upset or has any problem, we will talk it over and try to solve the problem. In solving the problem, we will never hurt, punish, criticize or put anyone down. We will always be willing to participate in a whole classroom discussion and anyone can ask for a class discussion at any time. In other words we can work out upsets, and problems by talking it over together with courtesy. **(Reference Bulletin #14)**

**Ask the students to generate suggestions for a class agreement about how we can solve problems in our classroom by working in Cooperative Learning groups. Illustrate, if appropriate to your grade level.**

6. Ask each Cooperative Learning group to present its suggested problem-solving agreement. Then the class develops **one** problem-solving agreement using a consensus process based upon ideas from all the groups.

✎ Record on Chart: **Work It Out**

7. Make posters of the agreement and post it in the classroom. Please record the agreement on the Staff Self-Evaluation and Reflections section, page 95.

## MY PLAN FOR APPLYING THE CONCEPTS
### . . . Students

8. This week when faced with problem-solving "opportunities" consider all the choices there are and how the problems could be solved courteously.

What is one thing I plan to do when I feel myself operating on my back two wheels? What is my plan? **Share plans.**

CHAPTER 12   Page 98

## APPLICATION OF LEARNING: Working It Out

9. Alice is a new student in our classroom and on her first day she begins to confront students in a way that is not courteous. **How can we help Alice move from upsetting to thinking and doing behavior? What basic need might Alice be meeting with her present behavior? Discuss together.**

## STUDENT PLANNING AND SELF-EVALUATION

10. **What was your plan to make better choices?** Think of a problem that happened this week at home, school or on the playground. What steps did you use to solve the problem effectively? Were you able to "work it out?" **Invite students to reflect in a journal, discuss with a friend or join in a class meeting/discussion as they evaluate their plans.**

Continue developing the idea of self-evaluation. This will lead students to make more responsible choices for themselves.

♪ *Notes & "Quotes"*

CHAPTER 12  Page 99

# JOURNEY TO QUALITY

## Journey 13
## The Environment for Non-Coercion

**DISCUSSION OUTCOMES AND LEARNINGS**

- to know the attributes of a friendly environment and ways of nurturing it
- to recognize that any coercion is detrimental to <u>maintaining</u> a friendly environment
- to understand that a non-coercive environment is basic for all relationships in a Quality School

The environment created in a classroom, staff room, school or a district directly affects how people treat each other within the organization. Creating a friendly environment is the first step in establishing a need-fulfilling classroom, school or district.

**A friendly environment is a place where people . . .**
- are treated with courtesy by caring staff members
- are provided opportunities to meet basic needs responsibly
- want to be because it is friendly
- perceive that problems can be worked out in non-threatening ways
- use persuasion rather than coercion as a way to influence others
- want to learn what is useful and will add quality to their lives
- accept we can only control ourselves

A non-coercive environment creates the foundation for building all other relationships within the school. It is our perception of whether we believe the environment is friendly or not that determines if it is need-fulfilling. In **Reference Bulletin #11**, Dr. Glasser writes that coercion decreases the quality of one's life. Even the slightest amount of coercion may create fear and threaten the relationship that has been established in the friendship environment.

**RECOMMENDED READINGS:**

Glasser, William
**The Quality School**, Chapter 3
**Bulletin #11, The Quality School Reference Bulletins**

# JOURNEY TO QUALITY

## DISCUSSION 13 WITH STAFF
### The Environment for Non-Coercion

### MAKING THE CONNECTIONS TOGETHER
### ... from Concepts to Practices

1. Visualize your favorite place to shop. What about that store makes you want to go there to shop and come back time and time again? **Think first by yourself, then share the specific qualities of your favorite shopping place with the group. Chart.**

     Record on Chart: **Shopping at the Best**

**Identify what basic needs are met when shopping at this store. Code each quality on the chart with an abbreviation of the basic needs it meets.**

2. **Think about and discuss briefly:**
   - Whose job is it to create the friendly environment?
   - How does an environment become friendly?
   - Why do some environments feel unfriendly?
   - Are there some offices or stores you refuse to patronize because they are not friendly?
   - How do you feel as you go in these places?

3. If we want to be in a friendly environment when we shop, doesn't it make sense that all of us would want to be in one at school? Without a friendly environment we cannot effectively manage, counsel or teach because this is the basis for all of our relationships and all that we do.

    If we can identify the qualities/attributes of a friendly environment in a place we like to shop, how many of these qualities/attributes do we have in effect in our school? **Circle those items listed on the chart, Shopping at the Best, we want in our school.**

4. Who is responsible for creating, nurturing and maintaining the environment in a Quality School? If it is the staff's role to create, nurture and maintain the environment, then what implication does this have for **each** of us? **Discuss and chart ways we can nurture a friendly Quality School environment.**

✎ Record on Chart: **Ways to Nurture a Friendly Environment**

5. As we look at the ways we've identified to nurture our school environment, what can we agree to do this week as a staff that will lead toward a more friendly atmosphere? Discuss and agree on a plan. Post in staff room.

✎ Record on Chart: **Our Plan For A More Friendly Environment**

6. Record the Staff Plan here:

## WORKING IT OUT WITH STAFF

7. John is the school "complainer." He gripes about everything: parents, students, fellow teachers, the building, the district and the world. He spreads his "cloud of gloom and doom" everywhere he goes. He has seen it all, heard it all and has a reason why every new suggestion won't work.

   ▸ What do we know about John's needs?
   ▸ Can we change him? Why? Why not?
   ▸ What are ways we can influence John?

   **Discuss as a staff.**

## MY PLAN FOR APPLYING THE CONCEPTS . . . Staff

8. Begin evaluating the environment of the places you go, stores where you shop, and meetings you attend during the week. Identify the things that enhance the environment and those that detract from it. Write what you observe on your Staff Planning and Self-Evaluation, page 104 at the end of the week. Be thinking of one thing you observed this week that you will plan on using to enhance the environment in your classroom.

## FROM THE STAFF ROOM TO THE CLASSROOM . . .

Once we gain new knowledge and understanding of the importance that the environment can have in meeting our basic needs, we quickly see the connection it has to all relationships within the school. Yet with this new knowledge we may still experience some personal struggles as we give up coercive practices. Take a few minutes during the week to ponder these questions from your new perceptions:

- What's wrong with a **little** coercion?
- How can a non-coercive person act effectively in a coercive world?
- How can we give up coercion without feeling like a doormat?

**Maintaining a friendly environment is much like nurturing a garden; once planted it must be cared for to keep it lovely.**

♪ *Notes & "Quotes"*

# STAFF: MY PERSONAL PLAN
## The Environment for Non-Coercion

My plan to apply the concepts personally:

◘ This week as I evaluated the environment of the places I went, I observed . . .

◘ One thing I **plan** to do next week to enhance the environment in my classroom is: (Record plan)

# STAFF SELF-EVALUATIONS AND REFLECTIONS . . .

◘ **Where I am** compared with **where I want to be** in creating and maintaining a friendly environment in my classroom.

◘ How do I rate the environment of my classroom on a scale of 1 to 10?

```
1 ----------------------------------- 5 ------------------------------------10
non-friendly                                                          friendly
```

Where do I **want** the environment in my classroom to be on the scale? What am **I** willing to do to get more of what I want?

◘ On a scale of 1 to 10 where do I place the environment in our school?

```
1 ----------------------------------- 5 ------------------------------------10
non-friendly                                                          friendly
```

Where do **I want** the environment in our school to be on the scale? What am **I** willing to do to get more of what I want?

# SUCCESSES On My Journey to Quality I Want to Share With Others:

# STAFF OBSERVATIONS 13   The Environment for Non-Coercion

1. On my journey to quality what did I do that worked well this week?

2. What would I do differently next time?

3. What is my biggest concern? What help do I need? What will I do?

## Ideas I Developed for Nurturing the Environment . . .

**Bring your journal to share at our next Journey To Quality staff discussion.**

# THIS WEEK IN THE CLASSROOM

## OUTCOMES  The Environment for Non-Coercion

- to know the attributes of a friendly environment and ways of nurturing it
- to recognize that any coercion is detrimental to <u>maintaining</u> a friendly environment
- to understand that a non-coercive environment is basic for all relationships in a Quality School

## GETTING STARTED WITH STUDENTS

1. Think back to the beginning of our class today and some of the things that the teacher did or said that felt friendly to you and made you glad that you are in our classroom. **Discuss specific things you remember in large group.**

2. Everyone in school is responsible for creating and maintaining the friendly environment. Think about things you did or said to other students or the teacher that were friendly? What did you do or say today? **Share with group.**

3. What are the things that students and teachers could do to make this a more friendly environment for all of us? **Think first alone. Share ideas as they are recorded.**

    ✎ Record on Chart: **Creating a Friendly Place To Be**

4. If it is important that we do all we can to make our classroom a more friendly place where we all want to be, what can we **plan** to do this week from our list that will help us? What can we all agree to do? **Discuss as a class and agree on a specific plan that can be accomplished this week. Post.**

    ✎ Record on Chart: **Our Plan for Creating a Friendly Classroom**

CHAPTER 13    Page 106

5. How is a day in our classroom different when we have a substitute teacher? Compare and contrast the classroom environment. How is it alike? How is it different? Who chooses how we behave when we have a substitute? **Discuss.**

6. List things we **would** see and **wouldn't** see people doing in a friendly environment. This activity can be a beginning step in identifying indicators of a friendly environment and assessing where we are in achieving it. **Discuss and record.**

## MY PLAN FOR APPLYING THE CONCEPTS . . . Students

7. Think of things we can do this week to help create and maintain the friendly environment in our class. What is one thing I am willing to do? **Share in class discussion.**

## APPLICATION OF LEARNING: Working It Out

8. Shawna is often "out of sorts with the world" and comes into our classroom in the morning feeling angry, upset and unhappy with herself and everyone.

   What happens to the friendly environment when one person's needs are not met? What can we do, as a class, to keep this a friendly place to be when one of our classmates has a problem? How can we use our agreement for **"work it out"** to help us? **Discuss.**

## STUDENT PLANNING AND SELF-EVALUATION

9. Record in your journal or draw a picture showing one way to create a friendly environment at school this week.

   ▶ What did I do this week to help our classroom be a friendly place?
   ▶ How do I feel when I'm in our classroom?
   ▶ What will I need to do differently in our classroom to maintain the friendly environment?

# JOURNEY TO QUALITY

## Journey 14
## Problem Solving Without Coercion

**DISCUSSION OUTCOMES AND LEARNINGS**

- to understand that if we are going to problem solve without coercion it may take a change in our Total Behaviors
- to understand that strategies to problem solve without coercion will depend upon the severity of the problem and the number of students involved
- to understand that maintaining a friendly relationship is basic to problem solving without coercion

The teacher and the student can mutually satisfy their need for belonging through friendship. Dr. Glasser writes, "Each (person) attempts to help the other find satisfaction in what they do together." **(Reference Bulletin #1)**

Learning and understanding that we control only ourselves is a shift in our thinking about how we have believed people behave. Most of us were taught by classroom methods that attempted to control us. Control Theory teaches us that our brain is an internal control system which drives our behavior. Teachers can use knowledge, persuasion, and establishing a friendly relationship with students as ways of influencing and problem solving.

Behavior incidents are a part of normal classroom interactions and cannot be avoided, according to Dr. Glasser. It is the way we agree to solve problems without coercion that is the main point to consider. In the teacher's role as counselor, effective strategies for solving problems use the Reality Therapy process of communication. It may be as brief as a 30-second intervention during teaching a lesson, a one-to-one conversation aside from teaching where both the teacher and the student focus on how to solve the problem, or as involved as a class meeting with all students in the class.

Learning the process of Reality Therapy requires training for effective use and implementation. The **Institute of Reality Therapy** (1-801-888-0688) offers training in Control Theory and Reality Therapy from certified instructors. Those in training soon learn that RT/CT is a self-improvement process **first**. Learning to control oneself is a beginning in learning to solve problems without coercion. Using the problem-solving agreement to "work it out," helps in maintaining a non-coercive environment.

**RECOMMENDED READINGS:**

Dr. Glasser, William
  **The Quality School**, Chapter 10
  **Bulletins # 1, 11, 14, The Quality School Reference Bulletins**

# JOURNEY TO QUALITY

## DISCUSSION 14 WITH STAFF
## Problem Solving Without Coercion

### MAKING THE CONNECTIONS TOGETHER
### . . . from Concepts to Practices

1. Think about an incident in your class when a student was so disruptive that you could no longer teach? What did you do? What did the other students do? Recreate the incident in your mind and think about **your** Total Behaviors. Did you notice any **physiological** effects? How were you **feeling**? What were you **thinking** and what did you **do**? (Remember that we always act as a Total Behavior System and all of our behaviors are happening at the same time.) **Process first by yourself, then share with the group.**

**Divide chart paper into four boxes; label each box with one of the total behaviors. Record a few examples of each behavior from the group.**

✎ Record on Chart: **Total Behaviors: Acting, Thinking, Feeling, Physiology**

2. Did behaving this way get you what you wanted when the disruption occurred?
    ▸ Did the student's behavior change?
    ▸ How did the other students act?
    ▸ Did the classroom environment or feeling tone change in any way that you noticed?

**Share with the large group after personally processing.**

3. What could you have done differently? Describe what you could choose to do the next time there is a disruptive incident. Think about each of the Total Behaviors.
    ▸ How can you keep the environment friendly and free of fear?
    ▸ What parts or "wheels" of your Total Behavior System will most help you solve problems non-coercively?
    ▸ What will you need to do in order to solve problems non-coercively in your classroom?

**Discuss together.**

4. How can thinking about the four parts of our Total Behaviors help us make better choices in solving problems with students? **Share briefly with someone sitting near you.**

## WORKING IT OUT WITH STAFF

5. Connie has taught in her building for 15 years and most parents perceive her as having excellent classroom management. Connie's management system is based upon a stimulus response approach form of discipline with increasingly more severe consequences. Suspension is the last alternative. A couple of the teachers in Connie's building believe her classroom is managed by fear and coercion. How can teachers influence Connie to consider another form of control in her classroom? How can a school eliminate coercive practices when some staff and parents may think they are working? **Discuss in a large group and share strategies of influence.**

## MY PLAN FOR APPLYING THE CONCEPTS . . . Staff

6. What is one thing I will plan to do this week that will help me maintain a friendly environment in my classroom when a behavior incident occurs? **Record on Staff Planning and Self-Evaluation page 114, now please.**

## FROM THE STAFF ROOM TO THE CLASSROOM . . .

In the past Reality Therapy may have been dismissed by some teachers who viewed it as a formal counseling session with one counselor and one client. They may have been quick to say, "I don't have that kind of time to deal one-on-one with a behavior incident when I have 28 other students."

Teachers with this perspective have not been willing to take a more in-depth look at Reality Therapy to see its broader application to the classroom in establishing the environment. For it is in this understanding that the greatest use can be made, especially when paired with Control Theory.

**Reality Therapy is a process of communication** that is made up of two parts
- the environment - where the teacher can act as friend and influencer
- the process - where the teacher's role is that of counselor who helps the student see that we can "work it out."

The teacher's role as a friend is something that can be developed and expanded all the time. It is the time spend in establishing a friendship that shortens the time needed for using the Reality Therapy process. With a resistive student the teacher may need to spend more than 90% of the time together building involvement and friendship (getting into the student's Quality World) and about 10% or less actually using the counseling process (Reality Therapy).

When behavior incidents in the classroom get in the way of quality teaching, the teacher may use the counselor role. Possibly a 30-second intervention using Reality Therapy will suffice to "work it out" or the teacher/counselor may need a longer time and will let the student know we will "work it out" later.

Dr. Glasser suggests in **Reference Bulletin #14** that teachers, skilled in using Reality Therapy, can counsel without taking more than a few moments away from their regular teaching activities. Neither teaching nor Reality Therapy will be effective until a friendly, non-coercive environment is established.

♬ *Notes & "Quotes"*

# STAFF: MY PERSONAL PLAN
## For Problem Solving Without Coercion

My plan to apply the concepts personally:

What is one thing I plan to do this week that will help me maintain a friendly relationship in my classroom when a behavior incident occurs? **Record plan here:**

---

## STAFF SELF-EVALUATIONS AND REFLECTIONS . . .

◘ How has the problem solving agreement, to "work it out," been helpful with students in my class?

◘ How well did I follow my plan for maintaining a friendly relationship in my classroom during a behavior incident this week? Share specifics.

◘ What improvement could I make on my part during a behavior incident this week? What could I do differently?

---

## SUCCESSES On My Journey to Quality I Want to Share With Others:

# STAFF OBSERVATIONS 14
## Problem Solving Without Coercion

1. On my journey to quality what did I do that worked well this week?

2. What would I do differently next time?

3. What is my biggest concern? What help do I need? What can I do?

# Ideas I Developed to Problem Solve Without Coercion . . .

**Bring your journal to share at our next Journey To Quality staff discussion.**

# THIS WEEK IN THE CLASSROOM

## OUTCOMES   Problem Solving Without Coercion

- ◘ to understand that if we are going to problem solve without coercion it may take a change in our Total Behaviors
- ◘ to understand that strategies to problem solve without coercion will depend upon the severity of the problem and the number of students involved
- ◘ to understand that maintaining a friendly relationship is basic to problem solving without coercion

## GETTING STARTED WITH STUDENTS

1. Think about a time you were in trouble. How did you solve the problem? **Discuss in large group. Describe the Total Behaviors.** (Remember the Total Behaviors car and the four wheels.)
    - ▸ What were you **feeling** at the time?
    - ▸ What was your **physiology**?
    - ▸ What were you **thinking**?
    - ▸ What were you **doing**?

♪ **NOTE TO THE TEACHER:** You may want to post a Total Behavior chart with the four wheels labeled or you may wish to have your students create their own Total Behavior car and label its four wheels. Older students might enjoy creating model cars to label with the Total Behaviors.

2. Solving problems in our relationships with other people is part of living and can always be worked out. Our classroom agreement to "work it out," can be helpful in solving problems in school as well as on the playground or at home, or any other places where problems develop. You are learning a valuable life skill because adults, just like students, need to learn to "work it out" when problems come about at work or with others at home, or with their friends.

Let's have some fun and role play (or discuss) some problems that could happen in school and practice using our class agreement to problem solve.

♪ **NOTE TO THE TEACHER:** You might ask Cooperative Learning groups to generate three or four classroom problems to use for role playing. Younger students could illustrate real problems. Encourage a variety of incidents ranging in severity to heighten interest in role playing.

Examples of role plays:

▸ A student is seen by the teacher while passing a note to another student.

▸ A student loudly accuses another student of taking something that belongs to him/her.

▸ An angry, defiant student refuses to do what the teacher asks and stomps out of the classroom.

Make cards for all role plays: teacher, acting out student, other students involved. Have students take turns playing parts, including role of the teacher.

3. You might want to role play the situations showing **effective** and **ineffective** strategies for problem-solving. Adapt the role plays to meet the skills and interests of your students and the amount of time you have to spend on this activity.

♪ *Notes & "Quotes"*

## MY PLAN FOR APPLYING THE CONCEPTS . . . Students

4. Discuss strategies for solving problems in a class meeting. During the week class discussions could include such questions as:

   - How am I doing in choosing to solve my own problems?
   - Am I using more "front wheel" or "back wheel" behaving?
   - How is my Total Behavior System operating?
   - Am I solving problems more effectively now?
   - What do I want to remember when a problem occurs that will help me?

## APPLICATION OF LEARNING: Working It Out

5. Create a **Problem Solving Jar** and ask students to write special problems they identify in the classroom, school, home or community. Draw several problems out of the Problem Solving Jar each day to use for role playing or in class meetings.

## STUDENT PLANNING AND SELF-EVALUATION

6. What is one thing I can do to help solve problems in our classroom?

What can I do to help keep the environment friendly in our classroom?

**Ask students to record in a journal or illustrate with pictures and key words. Follow up with a class meeting/discussion.**

# JOURNEY TO QUALITY

## Journey 15
## Using Self-Evaluation

> **DISCUSSION OUTCOMES AND LEARNINGS**
> - to understand the process of self-evaluation
> - to know how to use self-evaluation for improvement
> - to know the value of teaching self-evaluation to students
> - to understand that accepting the premise of internal locus of control is basic for using self-evaluation

Self-evaluation is the process of comparing the vision of what we want (Quality World) with the progress we are making toward getting it. We are measuring where we are with where we've been and where we want to be. **The Quality School**, which Dr. Glasser based on W. Edward Deming's concepts of quality, addresses self-evaluation as one of the key steps in creating a Quality School.

The process of self-evaluation uses questioning strategies from Reality Therapy and works effectively with both students and staff members. The self-evaluation questions are:

1. **What do I want?** (Quality World) This question helps create a clear vision about what I want and is a necessary first step in developing my Quality World picture of what I am willing to work for in my life. What are the indicators I would see if I had what I wanted?

2. **Where am I now in getting what I want?** (Perceived World) This question helps me compare where I am today with where I want to be.

3. **Am I getting closer to what I want?** (Comparing Place) This is **the** evaluation question and helps identify the gap between what I have and what I want based upon the indicators I identified. The gap sets off a signal (an urge to behave) for me to do something different. If I am not getting what I want, I might ask myself these questions: "Is what I'm doing getting me what I want? What else could I be doing? What help do I need?"

Without intentionally doing something differently, things will stay the same and the gap will remain. We've heard the old saying: If you keep doing what you've always done, you'll keep getting what you've always gotten.

4. **What will I do to get more of what I want?** (Total Behaviors) This question looks at a specific action I will take to get more of what I want. A more detailed look at this planning process follows in Chapter 16.

5. Dr. Glasser believes we are always self-evaluating, but we do not always act on the information. Most of the time people are willing to accept lower performances because they know it will take too much work to improve. The role of the teacher-manager is to establish a trusting relationship with the student/worker so that the students will come to believe that working hard will improve the quality of their lives. Over time the students learn to accurately evaluate their own work and look for ways to improve. This is the measure of a self-directed learner.

It may be helpful for staff members to use the self-evaluation process by applying it first to a personal goal such as an exercise program, a diet or improving a golf score.

Professionally, the same self-evaluation process can be used to help us achieve our vision of what we want as we progress toward quality in our classroom.

Teaching students to self-evaluate will help them make the move from wanting external evaluation of their work (grades, teacher praise) toward using internal evaluation, a prerequisite for quality.

Self-evaluation needs a risk-free atmosphere. In a friendly non-coercive, non-judgmental environment students will begin to learn that they have control over their own improvement just as teachers have learned the power of understanding internal locus of control, the basic tenant of Control Theory. It is really a self-control theory.

**RECOMMENDED READINGS:**

Glasser, William
**The Quality School, Chapter 11**
**Bulletins #2, 7, 12, Quality School Reference Bulletins**

# JOURNEY TO QUALITY

## DISCUSSION 15 WITH STAFF
## Using Self-Evaluation

### MAKING THE CONNECTIONS TOGETHER
### . . . from Concepts to Practices

1. Think about something new you've been trying to learn at home. Perhaps it is a new hobby, a foreign language you are learning before a vacation or a computer class. How do you know you are getting better? How do you personally evaluate your progress? **Think first by yourself. Then share ideas with someone sitting near you.**

2. In our personal life as well as in our professional life we have a desire for self-improvement. In an earlier discussion we created a vision (**Quality World**) of what we could become as a school. Are we getting closer to what we want (**Comparing Place**)? What are some indicators that could be used to evaluate where we are in our journey toward achieving our Quality School? (You may want to refer back to discussion chapter 7 in your journal.) **Return to the staff vision of our Quality School (page 51) and chart the indicators that are observable and can serve as measures of our school's progress.**

   ✎ Record on Chart: **What We Want/What We're Getting**

3. In a Quality School self-evaluation is possible in any area we want to improve, such as: How am **I** doing personally? How am **I** doing as a team member? How am **I** doing as part of a staff?

Begin using the Steps for Self-Evaluation on a **personal** level. Focus on one thing you want to improve on or get better at doing now. Try to be as specific as possible in selecting something that will be measurable and simple enough for the improvement to be noticeable/attainable in the short time frame of a week. Examples: reading for enjoyment half an hour daily, to eating healthy meals, planning to do something daily with your family.

## Steps for Self-Evaluation

A. **What do I want?**

   ▸ What is my vision of something I want to achieve? (personally or professionally)

   ▸ What will it look like when I get what I want? Describe the picture of what I will see in detail.

B. **Where am I now in getting what I want?**

   ▸ How close I am now in getting what I want? On a scale of 1-10 where am I?

C. **Is what I'm doing getting me what I want?**

   ▸ Am I closing the gap between what I want and what I have?

D. **What will I do now to get (more of) what I want?**

   ▸ What will I do now to make this happen?

   ▸ What are the actions I will do to get what I want? What is my plan? (**s**imple, **a**ttainable, **m**easurable and **s**pecific)

4. Read the steps for self-evaluation and complete steps 1, 2 and 3. Share your vision and the indicators of what you want with people sitting near you. **Discuss. Continue to complete the self-evaluation steps daily this week.**

## WORKING IT OUT WITH STAFF

5. Barbara is in her second-year of teaching and is beginning to develop a good understanding about her role as a teacher. She is warm and caring with students and expresses concern over their problems. She takes weekend classes, and purchases many materials to use in her classroom, but she has this nagging feeling that it's not enough. She often feels she'll never "get it together" in her classroom. It seems overwhelming. After school, Barbara often shares her frustrations with her teammates. She is trying too hard to please her teammates, her principal and parents of her students, but she is not pleasing herself. How can Barbara set reasonable expectations for herself? How can her teammates provide support? Who can really help Barbara? **Discuss with your colleagues.**

## MY PLAN FOR APPLYING THE CONCEPTS . . . Staff

6. What is my plan for self-evaluation this week? **Record on Staff Self-Evaluation page 123, and complete the Steps for Self-Evaluation during the week (page 121).**

## FROM THE STAFF ROOM TO THE CLASSROOM . . .

Self-evaluation helps both students and teachers take personal responsibility for their own growth. This is another way of saying that all of us are internally motivated, and we can choose to change our behavior at any time we want to change.

Taking responsibility for evaluating our own growth and improvement is an excellent way we can help ourselves progress on our personal journey to quality.

In **The Quality School**, Dr. Glasser points out that self-evaluation may be a new concept for both teachers and students and may be difficult at first. The journey to become a quality school, quality teacher and quality teammate is a five-year process that takes a lifetime.

## STAFF: MY PERSONAL PLAN
### Using Self-Evaluation

My plan to apply the concepts personally:

What is my plan for self-evaluation this week? How will I know I'm getting better? **Record plan here:**

---

## STAFF SELF-EVALUATIONS AND REFLECTIONS . . .

◻ Thinking back over the Chapter 15 discussion with other staff members, a helpful part of the discussion was:

◻ How well was I able to introduce my students to the self-evaluation process this week? What will I do next week?

◻ What did I discover about myself this week as I used the Steps to Self-Evaluation?

---

## SUCCESSES On My Journey to Quality I Want to Share With Others:

# STAFF OBSERVATIONS 15  Using Self-Evaluation

1. On my journey to quality what did I do that worked well this week?


2. What would I do differently next time?


3. What is my biggest concern?  What help do I need?  What can I do?


## Self-Evaluation Ideas I Tried . . .











Bring your journal to share at our next Journey To Quality staff discussion.

# THIS WEEK IN THE CLASSROOM

## OUTCOMES  Using Self-Evaluation

- to understand the process of self-evaluation
- to know how to use self-evaluation for improvement
- to know the value of teaching self-evaluation to students
- to understand that accepting the premise of internal locus of control is basic for using self-evaluation

## GETTING STARTED WITH STUDENTS

1. Think about a question your parents often ask you, "How are you doing in school?"
    - How do you know what to tell your parents?
    - Do you need a report card to give you information or do you already know how you are doing?
    - Is what you tell your parents sometimes different than what you know is true?
    - How do you measure your success?

   **Share with the group. Discuss and chart the students' indicators of success.**

   ✎ Record on Chart: **Ways I Know I'm Successful**

2. When sports stars want to improve their skills they often keep records of their skill improvement. They keep statistics of how they are doing so that they can tell if they are getting better. They use a process of **self-evaluation** to determine if they are getting closer to what they want for themselves. They compare where they are now with where they want to be and decide what to do to improve. This self-evaluation process helps them get more of what they want. **Knowing what you want (Quality World) is the first step in getting better.** Think about something you want to get better at doing during this week at school. It could be in a subject area, a sport, or something fun. Picture what you **want** (Quality World vision) very clearly. **Share in a small group.**

If you want to improve and get better at something, just like sports stars you will need
- to know where you are
- where you want to be
- ways to tell if you are getting better

If you don't do something differently, things will probably stay the same.

4. This week in our classroom we will learn a process that will help us get more of what we want. It's called **Self-Evaluation**. This will be useful for improving anything that you want to get better at doing.

## Using the Steps for Self-Evaluation With Students

a. **What do I want?**
- What is my vision of something I want to achieve? (at home or at school)
- What will it look like when I get what I want? Describe in detail the picture of what I will see.

b. **Where am I now in getting what I want?**
- How close I am now in getting what I want? On a scale of 1-10 where am I?

c. **Is what I'm doing getting me what I want?**
- Am I closing the gap between what I want and what I have?

d. **What will I do now to get (more of) what I want?**
- What will I do now to make this happen?
- What are the actions I will do to get what I want? What is my plan? (simple, attainable, measurable and specific)

**For teachers of younger students:** You may want to use a bar graph or design your own measurement instrument for helping your students chart their progress toward their goals.

♪ **NOTE TO THE TEACHER:** The Self-Evaluation process may be learned best in a step-by-step guided-practice approach. Each question should be asked, discussed and completed before going on to the next one. The pacing and amount of guided practice needed will depend upon the ages and readiness of the students.

## MY PLAN FOR APPLYING THE CONCEPTS . . . Students

5.  Provide time daily during the week to have students self-evaluate their progress toward their goal. They should reread their plan, chart their progress and decide if they need to make changes in their Total Behaviors. Review Chapter 11 for information on Total Behaviors and the example of the car.

## APPLICATION OF LEARNING: Working It Out

6.  Ron wants to improve at skateboarding but every time he tries to do the "tricky stuff" he falls off. So he throws the board, tosses it into the garage, and gives up in disgust. Knowing what you know about **self-evaluation,** what could Ron do differently? **Discuss together. Select other situations as desired.**

## STUDENT PLANNING AND SELF-EVALUATION

7.  Did you meet your goal for improvement this week?

    ▸ How did the self-evaluation process help you get more of what you wanted?
    ▸ Did you change your behaviors? (Total Behaviors) What did you do differently?
    ▸ What did you learn about yourself this week?

*♪ Notes & "Quotes"*

# JOURNEY TO QUALITY

## Journey 16
### Planning the Plan

---

**DISCUSSION OUTCOMES AND LEARNINGS**

- to understand that self-evaluation is the internal processing before making a plan for moving toward quality
- to understand planning is essential for ongoing improvement toward quality
- to analyze what's getting in the way of what we want and to recognize the conditions we can control as we plan

---

Self-evaluation is the process of comparing the vision of what we want (Quality World) with the progress we are making toward getting it. It is the internal process of measuring our progress or our plan for moving toward the quality we want.

When we recognize that we are not getting what we want we may need to replan by analyzing what's working and what isn't working. Closing the gap between what we want and what we are getting is an important part of the self-evaluation process. When we answer "no" to the key evaluation question, "Am I getting closer to what I want?" it will be necessary to replan. When the internal feedback tells us we are no longer making ongoing improvement toward closing the gap, replanning is a way to analyze the conditions we can control, think through our Total Behaviors and do something different.

Rethinking and replanning the plan can be viewed as an opportunity to fine tune and make necessary adjustments based upon our new perceptions and knowledge after using the plan.

**Replanning the plan** follows and builds upon the **Steps for Self-Evaluation**. It is included on the next page for your review.

## Steps for Self-Evaluation

A. What do I want?
B. Where am I now in getting what I want?
C. Am I getting closer to what I want?
D. What will I do to get more of what I want?

## Replanning the Plan

E. What do I want that I'm not getting?
F. What do I know now that I didn't know when I made my plan?
G. What is my plan? What will I commit to do now to get more of what I want?

**This is an expansion of the Replanning the Plan questions:**

E. **What do I want that I'm not getting?** This is revalidating what we originally thought we wanted. Is our picture of what we want clear enough to guide our behaviors? The more clearly defined the vision the more compelling it will be.

> ▶ What basic needs are not being met? Understanding what needs will be met will help us make a more responsible plan because basic needs drive all our behaviors. Choosing need-fulfilling behaviors increases the probability of a plan's success.

F. **What do I know now that I didn't know when I made my plan?** What do you know now as a result of trying the plan that gives you additional information?

> ▶ What is/isn't working? What parts of my original plan are working? This question helps identify what successes we can build upon and what needs adjusting. Closer analysis of a plan that isn't working helps us recognize that some parts of the plan may have been successful, thus validating the worth of the original plan. What parts of my plan aren't working and need replanning? Keeping the original plan in focus allows us to use the information we have learned as a resource for planning.

> ▶ What's getting in the way of my plan working? This is the time to brainstorm all the possible things that can be getting in the way of the plan's success. What are the roadblocks, events, beliefs, conditions, conflicting needs, knowledge, issues and people that can be barriers to the plan?

▶ What are the conditions I can/can't control? What choices of behavior do I have? On the brainstormed list above there will be some things I can control and some things I can't control. For example, I can control my own behaviors, but I can't control what another person does. When I identify a condition I can control, I can think through the choices of my behaviors and do something differently. When I recognize a condition over which I have no control, I stop trying to control it and focus my energies into those things that I can control. Taking effective control means choosing behaviors relating to the conditions I can control as opposed to blaming others or upsetting over things I can't control.

▶ Which behavior is the most responsible action for getting more of what I want? From the list of conditions I can control, what choices of behavior would be more need-fulfilling, responsible and get me more of what I want? Answering these questions will help me make the best choice of behavior I can make given what I know now.

G. **What is my new plan? What will I commit to do now to get more of what I want?** Without commitment, there is little chance of success for the plan. Will this choice compel me to action? When will I get started? How will I know I am making progress toward what I want? What will be the indicators that I am closing the gap between what I want and what I am getting. What evidence will I see? What actions will I take each day to get more of what I want? How will I record my progress? When will I self-evaluate by asking "Am I getting what I want by using this plan?" Do I want to continue with this plan? Do I want to make a new plan or replan?

Replanning provides an opportunity to refocus our actions as new information becomes apparent. Analyzing our plan offers continuous feedback and identifies conditions that may need to be adjusted to reach the identified goals. Continuing to work at what is and isn't working and fine tuning along the way is the process of self-evaluation at its best. This view of replanning, as a cyclical part of the process of getting what we want, is a hopeful outlook of life planning as we take responsibility for our own personal growth toward quality. As we work to become a quality school, students and staff can use planning and replanning as a process for moving toward more quality work and quality behavior.

The replanning questions follow the Reality Therapy process and can be used by the teacher effectively with both class replanning and individual student replanning.

**RECOMMENDED READINGS:**

Glasser, William
**The Quality School**, Chapter 11
**Bulletins #2 & 7, The Quality School Reference Bulletins**

# JOURNEY TO QUALITY

## DISCUSSION 16 WITH STAFF
### Replanning the Plan

**MAKING THE CONNECTIONS TOGETHER**
**. . . from Concepts to Practices**

1. Think about a time you made a special plan for the weekend and had anticipated carrying out the plan all week only to find at the last minute your plan wouldn't work. **Take a moment to write your answers, then briefly share with someone sitting near you.**
   - What was your plan?
   - What got in the way?

2. When a plan isn't working there are conditions or things within our control and conditions or things outside of our control. Focusing on those conditions within our control is a more effective way of dealing with the disappointment of not getting what we want. **Return to your plan in #1 and work with the same partner as you discuss the conditions within your control and the conditions outside of your control.**

3. What generalizations can you make about the conditions that can effect planning? **Share ideas in large group.**

4. Any time we have a plan that isn't working we can begin by looking at the conditions we can control. Replanning is a positive way of looking at any situation that isn't getting us what we want. Think about a plan for solving a current school problem that isn't working. **Brainstorm ideas and agree on an existing plan that needs replanning. Choose one that is possible to replan during this discussion time.** Work through the replanning process as a staff. See page 128 for a more detailed explanation. (The purpose of this activity is to work through the replanning process on a current school issue. Remember SAMS and keep it simple by selecting a less complex problem/plan.)

*CHAPTER 16   Page 131*

| **Replanning the Plan** | To follow the Steps for Self-Evaluation |

E. **What do I want that I'm not getting?** Is our picture of what we want a clear enough vision to guide our behaviors? What basic needs are not being met?

F. **What do I know now that I didn't know when I made my plan?** What is/isn't working? What's getting in the way of my plan working? What are the conditions I can/can't control? What choices of behavior do I have? Which behavior is the most responsible action for getting more of what I want?

G. **What is my new plan? What will I commit to do now to get more of what I want?**

---

5. How would this Replanning the Plan process help us as a staff on our Journey to quality? When would this be useful? How can we make the time for quality planning and replanning which is part of the self-evaluation process? **Discuss.**

6. What are some issues that staff members, team members and individuals might choose to work on with this replanning process? **Discuss with your team or staff and form problem solving groups as needed.**

## WORKING IT OUT WITH STAFF

A Social Studies Team developed a plan for writing curriculum units for their grade level. Each teacher accepted responsibility for completing a part of the unit. The day to put it together arrived. When the team sat down to work Donna said, "We had family things to do this week-end, and I just didn't get my part done." The unit is scheduled to begin in two days. How will the team work it out? How can the team replan so that everyone's needs are met? **Discuss.**

## MY PLAN FOR APPLYING THE CONCEPTS
### . . . Staff

What is my part in the staff replan (#4, page 131)? What will I do to meet my commitment that will help all of us get more of what we want? **Record on the Staff Planning and Self-Evaluation Page 134, now please.**

*CHAPTER 16    Page 132*

# FROM THE STAFF ROOM TO THE CLASSROOM...

Often we think of a plan as being final. It either works or it doesn't and then it's viewed as a failure. By changing our thinking and actions we can look at replanning as a time to make adjustments. Improvement is an indicator that we are making progress toward success. When we are not improving it's time to evaluate our plan. The self-evaluation question (Is what we are doing getting us what we want?) can be used personally, with teams or with staffs. The cyclical replanning process allows us to build on the successes of the initial plan and make adjustments as needed. The concept of replanning is a useful life skill to teach our students.

♪ *Notes & "Quotes"*

# STAFF: MY PERSONAL PLAN
## Replanning the Plan

My plan to apply the concepts personally:

What is my part in the staff replan? What will I do to meet my commitment that will help all of us get more of what we want? **Record plan here:**

## STAFF SELF-EVALUATIONS AND REFLECTIONS . . .

■ How did the concept of replanning offer new insights into problem solving:

■ What did I discover about myself during the staff replanning process for solving a school problem?

■ Did I use this process with a classroom plan that wasn't working? Describe and include the conditions that could be controlled and couldn't be controlled.

## SUCCESSES On My Journey to Quality I Want to Share With Others:

# STAFF OBSERVATIONS 16  Replanning the Plan

1. On my journey to quality what did I do that work well this week?

2. What would I do differently next time?

3. What is my biggest concern?  What help do I need?  What can I do?

# Replanning the Plan Ideas I Developed . . .

**Bring your journal to share at our next Journey to Quality staff discussion**

# THIS WEEK IN THE CLASSROOM

## OUTCOMES: Replanning the Plan

- to understand that self-evaluation is the internal processing before making a plan for moving toward quality
- to understand planning is essential for ongoing improvement toward quality
- to analyze what's getting in the way of what we want and to recognize the conditions we can control as we plan

1. Think about a time you and a friend made plans for something special and it didn't work out. Jot down or be ready to tell what happened. Share what you did when you got the news that you weren't going to be able to do what you had planned. **Describe your Total Behaviors.**

   ▸ What were you **feeling**?
   ▸ What was your **body** telling you?
   ▸ What were you **thinking**?
   ▸ What were you **doing**?

   Knowing that we always have a choice for how we behave, what else could you have done? Can you think of other choices you could have made? **Share with someone sitting near you, then discuss a few examples with the large group.**

2. Can you think of a time in our class when we made a plan and it worked well? What were the reasons it worked well? **Discuss.** Now can you think of a time in our class when our plan didn't work? **Brainstorm and select a plan that isn't working as well as it could be.**

   ✎ Record on Chart: **A Plan That Needs Work**

3. ♪ **NOTE TO THE TEACHER:** Here is a way you might talk about the replanning process with your students.

Sometimes our plans don't work out the way we expected them to work and it may be necessary to do something differently. Learning to evaluate our plans and making adjustments as we replan is a process that will help us be more successful. Replanning allows us to take advantage of what we have learned and gives us a foundation to build upon in replanning. It is important for us to recognize what's working in a plan and what isn't working so that we can make different, responsible choices. Using evaluation and replanning together is a way to continue to make improvement.

**Take the plan the class has selected to work on and use the replanning process to make adjustments.**

### Replanning the Plan
**To Follow Steps of Self-Evaluation A - D (Chapter 15)**

E. **What do I want that I'm not getting?** Is our picture of what we want a clear enough vision to guide our behaviors? What basic needs are not being met?

F. **What do I know now that I didn't know when I made my plan?** What is/isn't working? What parts of my original plan are working? What parts aren't working and need replanning? What's getting in the way of my plan working? What are the conditions I can/can't control? What choices of behavior do I have? Which behavior is the most responsible action for getting more of what I want? From the list of conditions I can control, what choices of behavior would be more need-fulfilling, responsible and get me more of what I want?

G. **What is my new plan? What will I commit to do now to get more of what I want?** When will I get started? How will I know I am making progress toward what I want?
**Discuss together in class discussion or meeting.**

4. The replanning process of rethinking and redoing our plans takes practice. What other classroom plans aren't working that need to be replanned? **Discuss and replan.**

### MY PLAN FOR APPLYING THE CONCEPTS
### .... Students

What is my part in the class plan? What will I do to help all of us get more of what we want? How will I record my progress each day?

## APPLICATION OF LEARNING: Working It Out

The class is divided into Cooperative Learning groups for a class project. Dana prefers to work alone and won't join the other members of her Cooperative Learning team in planning the project. The team members want to work together successfully. Dana is getting in the way by refusing to work with the group. What choices does the team have? What choices are more responsible? What could they decide to do? **Discuss.**

## STUDENT PLANNING AND SELF-EVALUATION

- Did I follow through on my part of our class plan?

- Did recording my progress each day help me keep my commitment?

**Record in your journal or illustrate with a picture or graph.**

# JOURNEY TO QUALITY

## Journey 17
## Problem Finding and Problem Solving: One Strategy for "Working It Out"

### DISCUSSION OUTCOMES AND LEARNINGS

- to recognize the teacher's role and the student's role in the problem solving process
- to recognize that problem finding and problem solving are a part of self-evaluation

Even with a friendly, non-coercive environment where thoughtful planning and replanning has taken place, problems will occur. They may happen at any time during the day. Having a way of managing them on an individual basis without interrupting the class or disturbing the environment is what every teacher wants. We want to use Reality Therapy as a way of helping our students learn to become self-directed, responsible citizens. Reality Therapy offers a way for maintaining an atmosphere in our classroom that is free of fear and open to the belief that we can always "work it out."

Taking the time to establish a procedure for how we can work on individual problems is a prerequisite for "working it out." Using the problem solving process gives the student an opportunity to think through the problem, their Total Behaviors, and make a plan as they self-evaluate. The student becomes an active participant in making responsible choices for how he will behave to get what he wants.

The "work it out" approach enables the student to accept ownership and choose more effective behaviors than when the teacher makes the plan for the student. Being willing to give the time that the problem solving process takes for thinking and planning requires a commitment from the teacher to allow the time and space for it to happen.

The belief that change is possible, and we can always learn to make more responsible choices is inherent in Control Theory and Reality Therapy. The teacher's role is to be open to helping students make more need-fulfilling plans for themselves. Some students may need more time and practice to learn more effective behaviors.

In the past, a coercive "quick fix" such as writing the student's name on the board or detention may have appeared to solve a problem temporarily. But the problem always reoccurred because the student took no personal ownership for his behavior. Working with a student to problem solve is time consuming, but in the long term the skill is well worth the time and effort it takes to develop.

There are several ways to make the process for problem solving and planning work in the classroom. Class meetings can serve as one way students can learn to problem solve. Students can also work individually to process problems using a teacher-made guide. Answering the problem-finding questions can lead naturally into planning for more effective behaviors. This process enables the teacher to keep teaching until there is time to meet with the student privately.

At other times a teacher may meet with small groups of students to resolve issues of concern.

Believing we can "work it out" and that together we can learn to manage our behavior more effectively sets the conditions for a quality learning environment. All of us in the school, both teachers and students, are learning we can only control ourselves. Learning Control Theory and applying Reality Therapy is new learning for all.

**RECOMMENDED READINGS:**

Glasser, William
**Bulletins #10 & 14, Quality School Reference Bulletins**

# JOURNEY TO QUALITY

## DISCUSSION 17 WITH STAFF
## Problem Finding and Problem Solving: One Strategy for "Working It Out"

### MAKING THE CONNECTIONS TOGETHER
#### .... from Concepts to Practices

1. Think about a student in your classroom who continually forgets to follow the class agreement for how we will treat each other. The student isn't choosing responsible behavior even after planning and replanning with the teacher. How might this situation be handled by a teacher using a coercive approach? **Think first by yourself and compare ideas with someone near you.**

2. At times a coercive method of handling problem behaviors appears to solve problems quickly, but this takes the responsibility away from the students and gives it to the teacher. " Quick external fixes" are not lasting. Students can be helped to become self-directed learners as they problem solve by taking responsibility for their own actions. This problem solving process requires time and patience for both the student and the teacher to learn.

    Understanding and accepting Control Theory means we believe we can all improve and that each individual has the capacity to make responsible choices. Change is possible for everyone.

    Look back to the problem behavior in #1, what are some non-coercive ways the teacher could have managed the situation differently? **Share in the large group.**

3. When a situation in the classroom gets in the way of learning, the teacher may ask the student to move to a designated area of the room to begin problem solving on his own. The student could write answers to the questions on the Problem Finding Guide to begin the process of planning for more responsible behavior while the teacher continues working with the class. (See Problem-Finding Guide on next page.)

    How could you use this Problem Finding Guide with students in your class? How would you organize and manage this? When would it be appropriate and useful? How would you follow up on the plan? **Divide by grade level teams for discussion. Report back to the large group and chart ideas.**

## Problem Finding Guide

Follows Replanning the Plan

H. What were you doing? Describe your Total Behaviors. How is what you are doing getting in the way of others in this classroom getting their needs met?

I. What will happen if you keep doing what you are doing? Is that what you want?

J. What do you believe is the problem now? Finding the problem may help with thinking through solutions. What do you know now that could help?

K. What do you want to do now to solve the problem?
 1. Talk to the teacher
 2. Make a plan
 3. Replan the Plan
 4. Identify the conditions you can and can't control
 5. Other

✎ Record on Chart: **Using the Problem Finding Guide**

4. Is the promise of lasting change worth the investment of time, energy and patience it will take to commit to applying Reality Therapy? Are you willing to use this idea? **Discuss with your team. Share with the large group.**

5. In order to begin, what help do you need? Would planning with a team be a support for you? Is there anyone else who could help? What would a plan for using the Problem Finding Guide with your students need to include? How would you set it up to use in your classroom? **Think first by yourself, then share your ideas.**

## WORKING IT OUT WITH STAFF

6. Mary Anne has been using the Problem Finding Guide in her classroom. This process is working well, but finding time to meet individually for follow up with the student is difficult. Mary Anne often chooses to meet with the student during her break. Although she is having success with her students, Mary Anne misses the interaction with her colleagues during the break. What can she do? **Discuss.**

## MY PLAN FOR APPLYING THE CONCEPTS . . . Staff

7. What is my plan for using the Problem Finding Guide with my students? What is one thing I will do this week to implement my plan? **Record on the Staff Planning and Self-Evaluation page 144 now, please.**

## FROM THE STAFF ROOM TO THE CLASSROOM . . .

The Problem Finding Guide is a specific "work it out" strategy for individuals. It provides a way for the teacher to handle an upsetting situation and continue teaching. Giving a student an opportunity to begin working it out provides the teacher and the student time for thinking before doing (working with the teacher). Completing such a guide should not be viewed as "punishment" if a need-fulfilling environment and trust have been established. This is a chance to use the class agreement, "We can work it out together." How this is done, (the attitude, body language, tone and manner) is critical for making this happen. The Problem Finding Guide is a first step in problem solving. It is not a replacement for the teacher and student interaction using Reality Therapy.

# STAFF: MY PERSONAL PLAN
## Using the Problem Finding Guide for Problem Solving

My plan to apply the concepts personally:

What is my plan for using the Problem Finding Guide with my class? What is one thing I will do this week to implement my plan? **Record plan here:**

---

## STAFF SELF-EVALUATIONS AND REFLECTIONS . . .

- How well did I follow my plan? **Share specifics.**

- What will I need to do to continue implementing this with my students?

- How could I use the Problem Finding Guide with the class to solve a class problem?

---

## SUCCESSES On My Journey to Quality I Want to Share With Others:

# STAFF OBSERVATIONS 17   Using the Problem Finding Guide for Problem Solving

1. On my journey to quality what did I do that worked well this week?

2. What would I do differently next time?

3. What is my biggest concern? What help do I need? What will I do?

# Ideas I Developed for Problem Finding and Problem Solving

Bring your journal to share at our next Journey To Quality staff discussion.

## THIS WEEK IN THE CLASSROOM

### OUTCOMES  Problem Finding and Problem Solving
- ◘ to recognize the teacher's role and the student's role in the problem solving process
- ◘ to recognize that problem finding and problem solving are a part of self-evaluation

## GETTING STARTED WITH STUDENTS

1. Think about a time you were in a Cooperative Learning group and one member of the group kept "goofing off" rather than doing the task assigned. How would you handle this situation? **Think first by yourself and then share in a group.**

2. Think about a situation during a lesson when the person sitting near you kept talking while the teacher was teaching. You couldn't hear the teacher and were concerned you might "get in trouble." What choices could you make? How might the teacher manage a student who disturbs others in the class? **Discuss and share ideas.**

3. In our class we have an agreement to "work it out." Sometimes people may choose not to follow the agreement, and it will be necessary to have them work out the problem. Often the teacher does not want to stop the class to work with one student's problem right then. The student could work on the problem alone with some helpful questions to guide the process until the teacher has time to meet with the student individually. Knowing everyone has problems and that problems are a part of life, learning a process for solving problems can be useful for everyone. What are some behavior problems that could happen in our classroom? **Brainstorm and chart.**

    ✎ Record on Chart: **Problem Behaviors in our Class**

4. Choose one behavior problem from the list to use with the Problem Finding Guide. Pretend you are a student who is having difficulty managing his own behavior. Let's practice using this Problem Finding Guide to learn to make more responsible choices.

CHAPTER 17    Page 146

Teach this step-by-step procedure so that all students understand and learn this process.

5. To provide independent practice, group the students into Cooperative Learning teams. Choose another problem behavior identified by the class and use the Problem Finding Guide to begin "working it out."

6. Why is it important to find and solve your own problems? **Discuss.**

## MY PLAN FOR APPLYING THE CONCEPTS . . . Students

7. Think about a problem you are having with another student or teacher. Use the Problem Finding Guide to begin to "work it out."

## APPLICATION OF LEARNING: Working It Out

8. John is always bothering you when you are trying to get your work done. The teacher does not notice the problem. What can you do? What are your options? How can you "work it out?" **Discuss.**

## STUDENT PLANNING AND SELF-EVALUATION

9. Did I practice using the Problem Finding Guide when a real problem occurred this week? What problem did I find? What did I do? **Write in your journal or illustrate.**

♪ *Notes & "Quotes"*

# JOURNEY TO QUALITY

## Journey 18
## Help Beyond the Classroom

**DISCUSSION OUTCOMES AND LEARNINGS**

- to develop a school plan for "working it out" outside the classroom
- to recognize that we will always need to work on resisting the use of coercive practices

In our ideal quality school, students and teachers would get their needs met in a responsible way, and help wouldn't be needed beyond the classroom for problems because there wouldn't be any. However, in the real world events take place that need more intervention than the classroom teacher can reasonably provide. Developing a building plan for working through problems both inside and outside the classroom is a proactive approach for a quality school. By looking at resources available, the staff can develop a plan for managing and counseling students who need help in solving problems. Most interventions can be managed within the classroom because the teacher and the students have come to an agreement on class rules and have created a willingness to work together to "work it out." At this classroom level the teacher uses the Reality Therapy process in helping students use self-evaluation feedback to develop their own plans for more effective behaviors.

Some problems require more time than the teacher can give and another staff member may be needed to provide a neutral perspective to the concern. For example, a student who is so disruptive that others cannot learn and is unwilling to "work it out" within the classroom will benefit from another level of intervention. A teacher may also need help with defiant students or with students who do not have pictures of school in their Quality World. In other words, we are working with students who are uncooperative and unwilling to plan at that time. It is important to identify who will be available to spend the time needed to influence and counsel the student to make more effective choices of behavior.

When problems are so severe that neither classroom nor counselor intervention is realistic, the principal will need to intervene. Drugs, weapons, assaults and property damage may be beyond the schools jurisdiction and society's laws will prevail. Dealing with very difficult students is the subject of Glasser's **Reference Bulletin #17**. There may be times when students choose such destructive behavior that they may not be in school until they are willing to make a plan for returning to school. Even at this severe level the school does not give up on the student or use coercion to solve the problem. Schools will continue to establish a need-fulfilling environment, influence, teach and persuade, but ultimately **the student is responsible for his own behavior.**

Our involvement with parents will require a shift in our thinking. In the past we may have informed parents when difficulties occurred believing they had a right to know. Often punishment was the result. Now we want students to learn to solve their own problems. Helping parents understand Control Theory and why we are eliminating coercion and fear is imperative as we work toward becoming a quality school. It is clear that schools will want to find ways of educating parents so that they understand our new way of thinking.

In order to develop a building plan for working out problems we will need to agree as a staff how help will be available, who will intervene and determine locations for working out problems. All people who work in the school will need an orientation on how we are going to work together to help students make responsible choices. Sharing our vision of working together and problem solving will help us develop consistency in how we treat each other. When the school cares, the students will care.

**RECOMMENDED READINGS:**

Glasser, William
**Bulletins #4, 10, 17, Quality School Reference Bulletins**

# JOURNEY TO QUALITY
## DISCUSSION 18 WITH STAFF
## Help Beyond the Classroom

### MAKING THE CONNECTIONS TOGETHER
### ... from Concepts to Practices

1. Just suppose that you could design the ideal "work it out" plan for the building. What would it look like? What would you see? What help and support would be available? **Think first by yourself and record the group ideas.**

    ✎ Record on Chart:  **Ideal "Work It Out" Plan**

2. What is our current school plan when a student needs more help in solving problems than the teacher\instructional assistant can give? **Discuss.**

3. What are the things we want for working out problems that we don't have right now? **Record on chart.**

    ✎ Record on Chart:  **What We Want?**

    What are the things on this list that are possible to implement without additional resources? **Circle the things on the list.**

4. What can we do now to develop a plan for working it out with our current resources? Develop a building plan. **Record.**

5. Are there additions we want to make to our plan that will require resources and time? How can we get what we want? **Discuss, plan and record.**

6. How will we communicate our building plan for working it out to new students, substitutes, volunteers; to all the people who come into our school? Could an audio visual presentation be developed to share with new people? What do parents need to know as we move along on our Journey to Quality eliminating fear and coercion? Would parent training be useful? Would this be something we would want to consider? **Discuss.**

## WORKING IT OUT WITH STAFF

7. Randi is an experienced teacher. This year her class has several volatile students. At any time pandemonium can break out. It seems the students feed off each other and all attempts at making plans for responsible behavior never last. The disruptive cycle continues daily. The teacher is discouraged and just wants the year to end. What are Randi's options in dealing with this difficult situation? How can her building staff offer her help and support? **Discuss.**

## MY PLAN FOR APPLYING THE CONCEPTS . . . Staff

8. What will I do to help develop a school plan for working it out? **Record on the Staff Planning and Self-Evaluation Page 152 now please.**

## FROM THE STAFF ROOM TO THE CLASSROOM . . .

9. Changing beliefs is difficult and the agreement to move from external locus of control to internal locus of control takes persistence and time and a willingness to be open to new ideas. When the staff works together it makes the transition easier because of the mutual support. We must be patient with ourselves and with our students as we learn more responsible ways of operating as a class. This is a 180 degree shift in our thinking from the teacher being in charge to everyone taking responsibility for their own action.

   Asking for help when it is needed will be part of the school plan. No one is left alone with a problem they cannot manage. Help is always available.

## STAFF: MY PERSONAL PLAN
### For Help Beyond the Classroom

My plan to apply the concepts personally;

What will I do to help develop a school plan for working it out? **Record plan here.**

## STAFF SELF-EVALUATIONS AND REFLECTIONS . . .

◘ How well did I follow my plan this week?

◘ The most difficult thing personally in eliminating coercion is.....

◘ What have I learned about eliminating coercion that would help others?

## SUCCESSES On My Journey to Quality I Want to Share With Others:

# STAFF OBSERVATIONS 18  Help Beyond the Classroom

1. On my journey to quality what did I do that worked well this week?

2. What would I do differently next time?

3. What is my biggest concern?  What help do I need?  What can I do?

# Ideas I Have Developed . . .

**Bring your journal to share at our next Journey To Quality staff discussion.**

# THIS WEEK IN THE CLASSROOM

### OUTCOMES  Help Beyond the Classroom

- to develop a school plan for "working it out" outside the classroom
- to recognize that we will always need to work on resisting the use of coercive practices

## GETTING STARTED WITH STUDENTS

1. What do you do when you need help at school? **Think first alone then discuss in a group.**

2. Who are the people in this school who can help? **Make a list of the people who could be helpful.**

    ✎ Record on Chart: **People Who Can Help Me**

3. What are some responsible and appropriate ways to get help when you need it? Contrast this to less responsible, inappropriate ways you can think of. **Discuss and role play if desired.**

4. Describe a situation when someone was afraid to ask for help? If keeping quiet isn't useful, what else could they do? **Discuss.**

5. How does our agreement for being courteous apply to situations when people we know need help? **Discuss in whole group.**

## MY PLAN FOR APPLYING THE CONCEPTS . . . Students

6. What are three things I can do when I need help? **List.**

7. What can I do when I ask for help and I am turned down? **Discuss.**

## APPLICATION OF LEARNING: Working It Out

8. Tara is harassed in the school cafeteria by three classmates as she tries to eat her lunch. First she ignores them, then she asks them to stop. Nothing works. What can Tara do? **Discuss.**

## STUDENT PLANNING AND SELF-EVALUATION

9. How am I doing in being able to solve my own problems and get help when I need it?

10. Ask students to journal about a time when they needed help and how they worked it out. Share your problem solving strategies with a partner.

# JOURNEY TO QUALITY

## Journey 19
## Making a Plan for Learning

### DISCUSSION OUTCOMES AND LEARNINGS
- to recognize that the concepts of self-evaluation and planning can be applied to learning
- to identify roles and responsibilities for learning

Learning is the common ground that connects everything we do in school and it is need-fulfilling. Learning provides each of us with a sense of accomplishment or power. It meets our need for fun through discovery and enjoyment and builds a sense of connectedness with other learners. We have the freedom to explore what interests and excites us. Quality learning doesn't happen without planning. Just as we teach strategies for solving problems for getting along with others we can help students learn strategies for solving learning problems. Students who learn quickly can also use the planning process to become self-directed as they extend their learning through personal interest investigations and research.

The self-evaluation questions can also be applied to learning: What do I want to learn? Where am I now in getting what I want? Is what I'm doing now getting me what I want? What will I do now to get what I want? The self-evaluation questions enable me to self-direct my own learning.

**If we want students to become self-directed in the pursuit of learning it is the teacher's role to see that the learning is meaningful, useful and will add quality to the students' lives.** It is the students' role to take responsibility for learning. Teaching students to use the process of self-evaluation will help them become more self-directed in the learning. When we combine the key components of **The Quality School**: teaching for quality, self-evaluation and eliminating coercion we provide the foundation for self-directed learning to flourish.

### RECOMMENDED READINGS:

Glasser, William
**The Quality School**, Chapter 8
**Bulletin #2, Quality School Reference Bulletins**

# JOURNEY TO QUALITY

## DISCUSSION 19 WITH STAFF
## Making a Plan for Learning

### MAKING THE CONNECTIONS TOGETHER
### ...from Concepts to Practices

1. Think about how renewed you feel each summer as you begin to look ahead to the fall and the start of school. Very often you attend a class and get a fresh idea about something you could add or do differently in your classroom. Once you are away from school you begin to see ways you could do things differently. How do you move from ideas to implementation in the classroom? Think about an idea you wanted to try this year? Did it happen? Did it work out as you pictured it might? **Share your ideas.**

2. How has meeting in a group for Journey to Quality helped us gain knowledge, ideas and support from each other? Have we gotten more of what we wanted by meeting together as a learning team? **Share how this journey has been helpful to you.**

3. We are now in Chapter 19 in **The Journey to Quality**. With one remaining chapter it is time for us to begin to think where we want to go from here with our learning. **Brainstorm a list of possible suggestions.**

   ✎ Record on Chart: **Our Next Learning Goal**

   Agree on one suggestion and begin to make a staff plan for continuing our Journey to Quality learning. **Record your staff plan.**

5. Will we continue to meet at the same time for support and learning? What is our plan? **Add to your staff plan.**

6. Who is responsible for my learning? How can the principal influence learning? What is the school's role, what is my team's role and what is my role? How can we help each other continue to grow? Would meeting in small learning teams be helpful? **Discuss in small groups. Share with the large group.**

## WORKING IT OUT WITH STAFF

7. Shirley has joined the staff second semester. How can we orient Shirley to the school's vision, journey and plan? Remembering that people may choose to join in our journey to quality when they are ready, would this be a good time to reissue the invitation? Is it important that everyone on the staff join? **Discuss and share ways to include everyone.**

## MY PLAN FOR APPLYING THE CONCEPTS . . . Staff

8. Think about something you want to learn personally or professionally. Use the following questions to make a plan for your learning.

    1. What do I want to learn?

    2. What will I do to begin learning what I want to learn?

    3. How will learning this get me what I want?

    4. When will I begin?

    **Record on the Staff Planning and Self Evaluation Page 159 now, please.**

## FROM THE STAFF ROOM TO THE CLASSROOM . . .

There is a difference between failing and having not yet learned. This is a change in our thinking from the idea that you learn it or you don't, to believing that with extra time learning is possible for all. As teachers, we never give up. It is the difference between having hope and losing all hope. If the learning isn't happening we can continue to plan and replan recognizing the conditions we can and can't control.

## STAFF: MY PERSONAL PLAN
### Making a Plan for Learning

My plan for applying the concepts personally:

What do I want to learn? What will I do to begin learning what I want to learn? How will learning this help me get what I want? When will I get started? **Record plan here.**

## STAFF SELF-EVALUATIONS AND REFLECTIONS . . .

- How well did I follow my plan for learning this week?

- Am I making the progress I thought I would make?

- How did making the commitment to a plan help?

## SUCCESSES On My Journey to Quality I Want to Share With Others:

# STAFF OBSERVATIONS 19  Making a Plan for Learning

1. On my journey to quality what did I do that worked well this week?

2. What would I do differently next time?

3. What is my biggest concern?  What help do I need?  What can I do?

# Making a Plan for Learning Ideas I Developed . . .

**Bring your journal to share at our next Journey To Quality staff discussion.**

# THIS WEEK IN THE CLASSROOM

## OUTCOMES  Making a Plan for Learning
- to recognize that the concepts of self-evaluation and planning can be applied to learning
- to identify roles and responsibilities for learning

## GETTING STARTED WITH STUDENTS

1. Think about all the things you have to learn in school this year? What is the hardest for you to learn and what is the easiest for you to learn? **Think first by yourself and share with someone near you? Extend into a class discussion.**

2. What do you do when something is hard for you to learn? How do you feel? What signals is your body sending you? What are you thinking and what do you usually do when something is difficult? **Share.**

3. When something is hard for you to learn, whose responsibility is it to see that you learn? What is your role in the learning? What is the teacher's role? Who can help you with the learning in school? Who can help you outside of school?

   ✎ Record on Chart: **Teacher's Role in Learning**

4. Have you ever thought something was too difficult for you to learn? Did you give up or did you finally learn it? **Share what you discovered about yourself as a learner.**

5. Sometimes when learning is difficult making a plan for learning is a way to get more of what you want? How can the process of planning be helpful? Think of something that is troublesome for you now. How could you use the planning process for learning? **Share in a large group.**

6. What is something we want to learn as a class this week? Let's make a class plan for learning.

A. What do we want to learn this week?
B. How will we know we are learning what we want to learn? What are the indicators we will see in order to know that we are learning?
C. What help will we need to learn this?
D. What will we need to do to help ourselves?
E. When will we begin?
F. What will we need to do differently if some people are learning and others aren't learning?
G. What is our plan?

✎ Record on Chart: **Our Class Plan for Learning**

## MY PLAN FOR APPLYING THE CONCEPTS . . . Students

7. After the class plan for learning is completed each student should be given the opportunity to personalize it. What will I do this week that will help me meet the class plan for learning? **Discuss your plan with the class.**

## APPLICATION OF LEARNING: Working It Out

8. Ricky doesn't finish or turn in any of his work. When asked about his work he often says he doesn't know what to do. What could the teacher do to help Ricky? What could Ricky do to take responsibility for his learning? **Discuss together.**

## STUDENT PLANNING AND SELF-EVALUATION

9. Did the class meet its plan for learning this week?

Did I meet my plan for learning this week?

How did making a plan for learning help me?

Would making a plan for learning each week be helpful?

# JOURNEY TO QUALITY

## Journey 20
## Continuing the Journey in a Quality School

### DISCUSSION OUTCOMES AND LEARNINGS

- to understand that the acceptance and use of Control Theory is basic for becoming a quality school
- to understand ways that change occurs

Where are we in our journey to quality as a school, as a team, as an individual? Periodically it is important for us to self-evaluate, recommit to the vision of a quality school and renew as we continue to move toward our vision of what we want.

It's time for us as a staff, teammates and individuals to evaluate our actions to see if what we are getting is what we want. Self-evaluation provides a process for feedback. Taking time to use the self-evaluation questions is valuable for giving us the knowledge that can lead to continued growth. This is the time for the staff to go back to the vision created in Chapter 7 to see where we are, where we have been and where we want to be. We will gain valuable information from the feedback process that will enable us to make adjustments or changes in our quest for quality.

Just as self-evaluation is useful for a staff to learn from feedback, it is equally helpful for individuals and teams. This chapter takes the staff through the Steps for Self-Evaluation together. At this time individuals and teams can recommit to their plan, make adjustments or create a new plan from the self-evaluation feedback.

How are we going to get more of what we want? Applying Control Theory with its concept of internal locus of control (we can only control ourselves) helps us understand how change occurs. One way change can happen is when we discover new ways we want to meet our basic needs. This is adding new pictures to our Quality World. We can also alter the pictures in our Quality World that we already have. For instance, a student may enter our school with learning being a very small part of his Quality World and school not meeting his basic needs. As we work to create a need-fulfilling environment for him, school may become a bigger picture in his Quality World. It becomes a place where he feels a sense of belonging, experiences fun and freedom and meets some of his need for power - provided his basic survival needs are met first.

Changing our Total Behaviors by doing something differently, is another way change can occur. Anytime we gather new information, gain insights or question our beliefs, we can change if it is need fulfilling for us.

Using the process of self-evaluation and understanding how change can occur offer us strategies for getting more of what we want. There will always be new staff members, new students and new parents who will be joining the journey in progress. Having a way for orienting these new people to the school's vision of quality and gaining their commitment to what we want is necessary in order for us to reach quality. Recommitting to the vision of a quality school is important for everyone who works in the school. By recommitting, a sense of renewal is felt by all.

**RECOMMENDED READINGS:**

Glasser, William
**Bulletins #1-17, Quality School Reference Bulletins**

# JOURNEY TO QUALITY

## DISCUSSION 20 WITH STAFF
## Continuing the Journey in a Quality School

### MAKING THE CONNECTIONS TOGETHER
### ...from Concepts to Practices

1. Think about the end of the day when the students have left and you are all alone and you have some quiet time to reflect. How do you evaluate the success of your day? What are the indicators you look for? What questions do you ask yourself? How do you get feedback? **Discuss and share with small group.**

2. Self-evaluation is a process for continual growth and improvement. It is how we give ourselves feedback. Dr. Glasser says we are constantly self-evaluating. The opportunity for growth and change is always available. As a staff, we can use the self-evaluation steps from Chapter 15 in a school evaluation to look at where we are in reaching our vision. Refer to the vision of a Quality School created in Chapter 7, page 51.

## *Evaluating Our Journey to Quality*

A. What do we want? Look at the indicators of a quality school we agreed upon. What is our vision of what we want? **Think for a few minutes alone and discuss.**

B. Where are we now in getting what we want? On a scale of 1-10 where are we now? **Discuss and share.**

C. Is what we are doing getting us what we want? **Are we getting closer?**

D. What will we do now to get more of what we want? What actions will we take next? Is there an area of focus we want to commit to now? **What is my plan?**

3. How often should we evaluate our journey to quality as a staff? **Process together and make a plan for future evaluation.**

4. Let's celebrate our journey toward becoming a quality school. Brainstorm our successes this year. Make a plan for where we want to go from here. A suggestion might be to consider using Glasser's three key points of a quality school, (eliminate coercion, teach for quality, and self-evaluate) as a way to look at successes and where we want to go from here.

## WORKING IT OUT WITH STAFF

5. When the staff members get together to discuss their plan for continuing their journey to quality next year, agree they want more time to work together as a staff. They would like to have a one hour early dismissal one day a week for quality school time. How can they influence others to make this happen? How will they work with the district? How will they involve the parents? How can they gain commitment to make their plan work? **Discuss and share plans.**

## MY PLAN FOR APPLYING THE CONCEPTS . . . Staff

6. What is one thing I plan to do this week to use the Steps of Self-Evaluation with students in my classroom? **Record on the Staff Planning and Self-Evaluation Page 167, now please**

## FROM THE STAFF ROOM TO THE CLASSROOM . . .

As you evaluated our school's journey to quality and your own journey, think about where you want to be and how you might plan to get there. What would help you get more of what you want personally? Perhaps you want to study Control Theory and practice the skills of Reality Therapy. Taking a Basic Week through the Institute of Reality Therapy may be helpful. What are more ways to implement the Quality School concepts in your classroom? Would processing your notes from **The Journey to Quality** help you gain new insights? Would developing a personal summer reading plan and sharing what you read with a teammate be fun? What would be the most helpful to you on your journey?

# STAFF: MY PERSONAL PLAN
## Continuing the Journey in a Quality School

My plan to apply the concepts personally:

What is one thing I plan to do this week to use the concept of self-evaluation with **my** students in my classroom? **Record plan here.**

---

## STAFF SELF-EVALUATIONS AND REFLECTIONS . . .

◘ How well did I follow my plan to teach my students to apply the concepts of self-evaluation?

◘ As I compare the student's self-evaluation of the year with my self-evaluation of the year, how close were our perceptions?

◘ What will I do differently after asking myself the self-evaluation questions?

---

## SUCCESSES On My Journey to Quality I Want to Share With Others:

# STAFF OBSERVATIONS 20
## Continuing The Journey to a Quality School

1. On my journey to quality what did I do that worked well this week?

2. What would I do differently next time?

3. What is my biggest concern? What help do I need? What can I do?

## Ideas I Developed as I Self-Evaluated My Journey to Quality

**Bring your journal to share at our next Journey To Quality staff discussion.**

# THIS WEEK IN THE CLASSROOM

## OUTCOMES Continuing The Journey in a Quality School
- to understand that the acceptance and use of Control Theory is basic for becoming a Quality School
- to understand ways that change occurs

## GETTING STARTED WITH STUDENTS

1. Think about this year in school and how much you have learned. How do you know how much you have learned? What can you do now that you couldn't do when you came into our class? **Discuss and share with the group.**

2. The answers in the first activity give us feed back on how we are doing. It's a way of knowing if we are making improvement and growth. In fact, Dr. Glasser says we are always processing how we are doing, and when we do this we have an opportunity to make changes based upon what we have learned. Think back to our plan for working for quality and the things we would do to help our class move toward quality. Let's use the self-evaluation questions as we compare where we are with where we said we wanted to be. (Look back to Chapter 7, page 55-56.)

### *Evaluating Our Journey to Quality*

A. What did we say we wanted in our Quality Classroom? What was our vision of what we said we wanted? **Discuss with the group.**

B. Where are we now in getting what we wanted? On a scale of 1-10 where are we now in getting what we wanted? How did we decide where we are? What indicators did we use? Work in Cooperative Learning groups. **Discuss and share.**

C. Is what we are doing as a class getting us what we wanted? Are we getting closer? **Discuss.**

D. What will we do now to get more of what we want? What actions will we take next to make this happen?

E. If we want to get more quality what can we plan to do? What is one thing we can do now to help us get more quality? What help do we need? **Discuss in small group.**

3. If you were going to fill out your report card what would you write? What are your strengths? In what areas are you improving? What comments would you write to yourself? **Discuss.**

4. How will you meet your basic needs this summer? What will you do to be sure you get enough belonging, power, freedom and fun? What do you want to learn this summer when you aren't in school? What help will you need? **Share in small groups.**

## MY PLAN FOR APPLYING THE CONCEPTS . . . Students

5. What is one thing I plan to do this week that will help our classroom get more of what we said we wanted? Share the plans in a class discussion.

## APPLICATION OF LEARNING: Working It Out

6. Molly is great at making a plan for changing her behavior but nothing different happens. Her behavior gets in the way of everyone's learning. Molly is often defiant and says, "You can't make me." How would learning about self-evaluation help her? **Discuss together.**

## STUDENT PLANNING AND SELF-EVALUATION

7. What did I do this week to meet my own basic needs and help my class get more quality in our classroom? How well did my plan work for me this week? Would making a plan every week help me get more quality? Who could I teach the self-evaluation process to this week?

# REFERENCES

Glasser, William. *Control Theory.* New York, Harper & Row, 1984.

Glasser, William. *Control Theory in the Classroom.* New York, Harper & Row, 1986.

Glasser, William. *The Quality School.* New York, Harper & Row, 1990.

Glasser, William. *Quality School Reference Bulletins.* Institute for Reality Therapy, Canoga Park, CA., 1991.


For more information on Reality Therapy/Control Theory contact:

The Institute for Reality Therapy
7301 Medical Center Drive, Suite 407
Canoga Park, CA 91307
Phone: (818) 888-0688
Fax: (818) 888-3023

## About the Authors

*Mona Perdue* and *Mariwyn Tinsley* have been friends and colleagues for the last 26 years and share a number of pictures in their Quality Worlds: sons, cats, foreign movies, opera, theater, travel, latte', walking, talking, and planning their next journey.

*The authors began their personal and professional journey while teaching kindergarten and found success collaborating on developing curriculum materials over the years. Mona and Mariwyn have taught in both the public school system and at the college level.*

*Currently both authors are certified Outcome Based Education Trainers and are planning to complete certification with the Institute of Reality Therapy in Vancouver, BC in 1992. They present workshops and serve as consultants to other school districts.*

*Mona Perdue has an MA in Early Childhood Education from the University of Washington and currently teaches 2nd grade at View Ridge Elementary School (a Quality School) in Bremerton, Washington.*

*Mariwyn Tinsley has an MA in Education and Administration from Seattle Pacific University and is currently the Administrator for Outcome Based Education/Curriculum for the Bremerton School District (a Quality District) in Bremerton, Washington.*